Beyond Infinity

Also by Eugenia Cheng

How to Bake Pi: An Edible Exploration of the Mathematics of Mathematics

Beyond Infinity

AN EXPEDITION TO THE OUTER LIMITS OF MATHEMATICS

Eugenia Cheng

BASIC BOOKS
New York

First published in the United States by Basic Books, an imprint of Perseus Books, LLC, a subsidiary of Hachette Book Group, Inc.

First published in Great Britain by Profile Books LTD, 3 Holford Yard, Bevin Way, London WC1X 9HD, www.profilebooks.com.

A catalog record is available from the Library of Congress.
Library of Congress Control Number: 2017931084
ISBN: 978-0-465-09481-3 (hardcover)
ISBN: 978-0-465-09482-0 (e-book)

10 9 8 7 6 5 4 3 2 1

In memory of Sara Al-Bader

who taught me by example that
infinite love can fit into a finite life.

Contents

Prologue

I hate airports.

I find airports stressful, crowded, noisy. There are usually too many people, too many queues, not enough seats, and unhealthy food everywhere tempting me to eat it. It's a shame when this is the way traveling starts, as it makes me dread the journey. Traveling should be an exciting process of discovery. Airports – along with cramped economy seating – too often mar what is the almost miraculous and magical process of flying somewhere in a plane.

Mathematics should also be an exciting process of discovery, an almost miraculous and magical journey. But it is too often marred by the way it starts, with too many facts or formulae being thrown at you, and stressful tests and unpalatable problems to solve.

By contrast, I love boat trips.

I love being out on the open water, feeling the wind in my face, watching civilization and the coastline recede into the distance. I like heading toward the horizon without it ever getting any closer. I like feeling some of the power of nature without being entirely at its mercy: I'm not a sailor, so usually someone else is in charge of the boat. Occasionally there are boats I can manage, and then the exertion is part of the reward: a little rowing boat that I once rowed around a small moat encircling a tiny chateau in France; a pedal boat along the canals of Amsterdam; punting on the river Cam, although after once falling in I was put off for life, just like some people are put off mathematics for life by bad early experiences. I have taken boat trips to see magnificent whales off the coast of Sydney and Los Angeles, or seals and other wildlife off the coast of Wales. Then there are the ferries crossing the Channel to France that started most of our family holidays when I was little, before the improbable Eurostar was built. How quickly we humans can come to take something for granted even though it previously seemed impossible!

These days I rarely take boats with the purpose of getting anywhere – rather, the purpose is to have fun, see some sights or some nature, and possibly exert myself. The one exception is the Thames River ferry, which is a very satisfying way to commute into central London, joyfully combining the fun of a boat trip and a journey with a destination.

I love abstract mathematics in somewhat the same way that I love boat trips. It's not just about getting to a destination for me. It's about the fun, the mental exertion, communing with mathematical nature and seeing the mathematical sights. This book is a journey into the mysterious and fantastic world of infinity and beyond. The sights we'll see are mind-boggling, breathtaking, and sometimes unbelievable. We will revel in the power of mathematics without being at its mercy, and we will head toward the horizon of human thinking without that horizon ever getting any closer.

part one

THE JOURNEY

1 What Is Infinity?

Infinity is a Loch Ness monster, capturing the imagination with its awe-inspiring size but elusive nature. Infinity is a dream, a vast fantasy world of endless time and space. Infinity is a dark forest with unexpected creatures, tangled thickets, and sudden rays of light breaking through. Infinity is a loop that springs open to reveal an endless spiral.

Our lives are finite, our brains are finite, our world is finite, but still we get glimpses of infinity around us. I grew up in a house with a fireplace and chimney in the middle, with all the rooms connected in a circle around it. This meant that my sister and I could chase each other round and round in circles forever, and it felt as if we had an infinite house. Loops make infinitely long journeys possible in a finite space, and they are used for racetracks and particle colliders, not just children chasing each other.

Later my mother taught me how to program on a Spectrum computer. I still smile involuntarily when I think about my favorite program:

```
10 PRINT "HELLO"
20 GOTO 10
```

This makes an endless loop – an abstract one rather than a physical one. I would hit `RUN` and feel delirious excitement at watching `HELLO` scroll down the screen, knowing it would keep going *forever* unless I stopped it. I was the kind of child who was not easily bored, so I could do this every day without ever feeling the urge to write more useful programs. Unfortunately this meant my programming skills never really developed; infinite patience has strange rewards.

The abstract loop of my tiny but vast program is made by the program going back on itself, and self-reference gives us other glimpses of infinity. Fractals are shapes built from copies of themselves, so if you zoom in on them they keep looking the same. For this to work, the detail has to keep going on "forever," whatever that means – certainly beyond what we can draw and beyond what our eye can see. Here are the first few stages of some fractal trees and the famous Sierpinski triangle.

If you point two mirrors at each other, you see not just your reflection, but the reflection of your reflection, and so on for as long as the angle of the mirrors permits. The reflections inside the reflections get smaller and smaller as they go on, and in theory they could go on "forever" like the fractals.

We get glimpses of infinity from loops and self-reference, but also from things getting smaller and smaller like the reflections in the mirror. Children might try to make their piece of cake last forever by only ever eating half of what's left. Or perhaps you're sharing cake, and everyone is too polite to have the last bite so they just keep taking half of whatever's left. I'm told that this has a name in Japanese: *enryo no katamari*, the last piece of food that everyone is too polite to eat.

We don't know if the universe is infinite, but I like staring up at a church spire and tricking myself into thinking that the sides are parallel and it's actually an infinite tower soaring up into the sky to infinity. Our lives are finite, but fictional and mythological tales of immortality appear through the ages and across cultures.

So much for glimpses of infinity, like ripples in the waters of Loch Ness that may or may not have been caused by a giant, ancient, mysterious monster. What is this monster we call infinity? What do we mean when we throw around the innocuous-sounding word "forever"? In our modern impatient world, people are prone to using it rather hyperbolically, exclaiming "I've been on hold *forever*!" after two minutes, or "This web page is taking *forever* to load!" if it takes more than about three seconds. I'm told by Basque writer Amaia Gabantxo that in Basque the word for eleven is *hamaika*, but it also means infinity. This is borne out by another friend of mine who did an audit of homemade jam in the cupboard and declared there were "four jars from 2013, ten from 2014, and lots from 2015." Apparently more than ten might as well be infinity. My research field is higher-dimensional category theory, and there "higher" usually means three dimensions or more, including infinity – everything from just three to infinity turns out to be about the same.

The way we think about infinity in our normal lives might be dreamy and exciting, but it dissipates under close examination, like the end of a rainbow, which will never be there if you try to go looking for it. It causes paradoxes and contradictions, impassable ravines and murky traps. It doesn't stand up to the tests of rigorous logic, as we'll see.

One of the roles of mathematics is to explain phenomena in the world around us, especially phenomena that crop up in many different places. If a similar idea relates to many different situations, mathematics swoops in and tries to find an overarching theory that unifies those situations and enables us to better understand the things they have in common. Infinity is one of those ideas. It pops up all over the place as an idea that we can dream about, and seems similar to other ideas that *can* be unified by mathematics: the ideas of length, size, quantity. So why is it so difficult to extend those easy mathematical concepts to include infinity? This is what this book is about: why it is difficult, how it eventually can be done, and what we see along the way.

Infinity Instincts

Infinity is easy to think about but hard to pin down. Small children can quickly latch on to the idea of infinity, but it took mathematicians thousands of years to work out how to explain infinity in full-technicolor logical glory. Here are some things we might think about infinity. Children come up with these thoughts about infinity quite often, all by themselves:

Infinity goes on forever.
Infinity is bigger than the biggest number.
Infinity is bigger than anything we can think of.
If you add one to infinity it's still infinity.
If you add infinity to infinity it's still infinity.
If you multiply infinity by infinity it's still infinity.

Children can get very excited when the notion of infinity first dawns on them. They learn to count up to ten, and then twenty, and then they learn about a hundred, a thousand, a million, a billion. If you ask a small child what the biggest possible number is, they may well say "a billion," but then you can ask them about a billion and one and watch their eyes widen.

It's not very hard to convince them that no matter what number they think of, you could always add one and get a bigger number. This gives the idea that there is *no biggest number*. Numbers go on forever! But then, how many numbers are there in total? The idea of infinity starts to emerge.

Perhaps some children first hear of infinity because they watch the *Toy Story* films and they hear Buzz Lightyear saying, "To infinity . . . and beyond!," which does sound very exciting. When I was a child *Toy Story* hadn't been made yet, but I had inklings of infinity from the loops I described earlier, the physical loops in our house and the abstract loops in my favorite computer program.

Once children start thinking about infinity, they can easily come up with questions about it that are teasingly difficult to answer. What is infinity? Is it a number? Is it a place? If it's not a place, how can we go there and beyond?

If children hear about infinity at school, the questions just start proliferating. Is one divided by zero infinity? Is one divided by infinity zero? If infinity plus one is infinity, what happens when you subtract infinity?

When children ask innocuous math questions that seem impossible to answer, this can be intimidating for adults, who feel that they are supposed to have all the answers. But as math educator and innovator Christopher Danielson says, an important aspect of learning is being able to ask new questions; this is more important than being able to state new facts. In math, there are always more questions. Even people who did quite well at math, even people who did math at university, even research mathematicians have more questions about infinity than can be answered.

Infinity Weirdness

Here are some of my favorite mind-boggling conundrums about infinity that we are going to explore.

- If you have an infinite hotel and it's full, you can still fit another guest in by moving everyone up one room.
- If there were a lottery with infinitely many balls, what would be your chances of winning?
- Some infinities are bigger than others!
- Infinite pairs of socks are somehow more infinite than infinite pairs of shoes.
- If I were immortal, I could procrastinate forever.
- If you are traveling from A to B, you first have to cover half the distance, then half the remaining distance, and then half the remaining distance, and so on. There will always be half the remaining distance left, so you'll never get there. Or will you?
- The recurring decimal $0.\dot{9}$ *is exactly equal to* 1.
- Does a circle have infinitely many sides?
- Why do people who do quite well at math often get stuck at calculus? Yes, this is a question about infinity too.

Infinity can appeal to people of all ages, of all levels of expertise, in different ways. This book is going to be a journey to infinity – and beyond. Because there really is something beyond infinity, if you think about it hard enough and in the right way. Just like there are always more questions to ask and more things to understand. Infinity is not a physical place, so this is not a physical journey. You can go with me on this journey sitting right where you are, because the journey is an *abstract* journey.

It's a journey into the deep, tangled, mysterious, murky, endless world of ideas.

Why?

Why do we go on this journey? Like with physical journeys, there are many reasons to go on an abstract journey. Everyone has their own reasons for traveling. Maybe there's something particular you want to do at your destination. Maybe there's a really good view from the top. Maybe there's beautiful scenery along the way. Maybe you enjoy the physical exertion of walking or climbing, or the exhilaration of riding a bicycle or driving a fast car, or the serenity of sitting on a train watching the countryside rush past. (My experience of trains involves more delays and irate commuters than serenity, but let's ignore that for a second.) Maybe you like going into the unknown. Maybe you like wandering around and losing yourself completely in a city. Maybe you have the travel bug and want to see as much of our incredible earth as possible, simply because there is so much to see.

All of these reasons have their counterparts in the abstract world. A journey where you have something particular to do (like commuting to work) corresponds to having a particular problem you want to solve or a particular application in mind. This type of abstract journey is less about the sense of discovery and more about getting something done. Going up high for a good view is like the abstract investigations we do to gain new perspectives on things we've already seen up close. The beautiful scenery along the way is the weird and wonderful ideas and scenarios that come up as we investigate. And yes, there is exhilaration in the mental exertion, and excitement in thinking about ideas that seem incomprehensible and then slowly clear up, like fog clearing to reveal the ocean glistening to the horizon. I don't have the travel bug for the physical world as much as some people do, but I do have the curiosity bug, the counterpart for the abstract world. I'm fairly calm and resigned about there being a lot of the world I haven't seen, but I am insatiable when it comes to ideas I don't understand. I always want to explore them. As soon as I catch a glimpse of something I don't understand, I am driven to plunge into it. I like losing myself completely in a city, and I like losing myself in ideas as well. As much as I am driven to understand things, I am also happy to acknowledge the things we humans *cannot* understand. In fact, I positively revel in it. It means that there is always more out there, which is a glorious thing. Wouldn't it be a bit sad if you could finally say yes, that's it, I've been to every restaurant in London now? But of course that's impossible. There will *always* be another restaurant you haven't tried, and there will *always* be things we don't understand.

In a strange way this book isn't about infinity at all. It's about the excitement of a journey into the abstract unknown. Jules Verne's *Journey to the Center of the Earth* wasn't really about the center of the earth, but about the excitement of an incredible journey. This is a book about how abstract thinking works and what it does for us. It's about how it helps us pin down what we really mean when we start having an interesting idea. It doesn't necessarily explain the whole idea: mathematics doesn't explain everything about infinity. But it does help us become clear about what we can and can't do with infinity.

This first part of the book, then, is a journey toward understanding what infinity is. If you ask a small child what they think infinity is, they might say something like "It's bigger than any number you can think of." This is true, but it still doesn't tell us what infinity *is*, any more than saying "Yao Ming is taller than anyone you've ever met" tells you who Yao Ming is.

In the second part of the book we'll take a tour of the world equipped with our new ideas about infinity and see where this elusive creature has been lurking all along. It's in mirrors that we point at each other, racetracks that we run around, every journey we take, and in every changing situation in this continuously changing world of ours. Understanding infinity is the basis of the field of calculus, which is inextricably embedded in almost every aspect of modern life.

Of course, it is possible to enjoy all those aspects of modern life without having the slightest understanding of calculus, which is why I don't emphasize these applications as the main reason for writing about infinity. Mathematics suffers a strange burden of being required to be useful. This is not a burden placed on poetry or music or football. If you ask me what all this is useful for, I could answer by saying it helps us generate electricity; make phone calls; build bridges, roads, and airplanes; provide water to cities; develop medicines; save lives. But this doesn't mean it is useful for *you* to think about it, only that it is useful to you that *somebody else* thought about it. It is not why I think about it, and not why I most want to tell this story.

You can get by perfectly well in life without understanding anything more about infinity than you did when you were five years old. But for me the usefulness of mathematics isn't about whether you need it to "get by" in life or not. It's about how mathematical thinking and mathematical investigation sheds light on our thought processes. It's about taking a step back from something to get a better overview. Flying higher up in the sky enables us to travel farther and faster.

Let's go.

2 Playing with Infinity

Mathematics can be thought of as many things: a language, a tool, a game. It might not seem like a game when you're trying to do your homework or pass an exam, but for me one of the most exciting parts of doing research is when you're just starting something new and you get to play around with some ideas for fun. It's a bit like playing around with ingredients in the kitchen, which is more fun than trying to write down the recipe you invented in case you want to repeat it. And that is in turn more fun than trying to write down the recipe for *someone else* to be able to repeat.

I'm going to start by playing around with the idea of infinity a bit, to free our brains up and start exploring what we think might be true about it and what the consequences are. Mathematics is all about using logic to understand things, and we'll find that if we're not careful about exactly what we mean by "infinity," then logic will take us to some very strange places that we didn't intend to go. Mathematicians start by playing around with ideas to get a feel for what might be possible, good and bad. When Lego was first made, the designers must have played around with some prototypes first before settling on the wonderful final design.

A mathematical "toy" should be like Lego: strong enough to be able to build things, but versatile enough to open up many possibilities. If we come up with a prototype for infinity that causes something important to collapse, then we have to go back to the drawing board. After our initial games we'll be going back to the drawing board several times as we move through different ways of thinking about infinity that go wrong and cause things to collapse. When we finally get to something that holds up, it might not look the way you were expecting. And it causes some things to happen that you might not have been expecting either, like the weird fact that there are different sizes of infinity, so that some things are "more infinite" than others. This is a beautiful aspect of any kind of journey – discovering things you weren't expecting.

In the previous chapter I listed some beginning ideas about infinity.

Infinity goes on forever.

Does this mean infinity is a type of time, or space? A length?

Infinity is bigger than the biggest number.
Infinity is bigger than anything we can think of.

Now infinity seems to be a type of size. Or is it something more abstract: a number, which we can then use to measure time, space, length, size, and indeed anything we want? Our next thoughts seem to treat infinity as if it is in fact a number.

If you add one to infinity it's still infinity.

This is saying

$$\infty + 1 = \infty$$

which might seem like a very basic principle about infinity. If infinity is the biggest thing there is, then adding one can't make it any bigger. Or can it? What if we then subtract infinity from both sides? If we use some familiar rules of cancellation, this will just get rid of the infinity on each side, leaving

$$1 = 0$$

which is a disaster. Something has evidently gone wrong. The next thought makes more things go wrong:

If you add infinity to infinity it's still infinity.

This seems to be saying

$$\infty + \infty = \infty$$

that is,

$$2\infty = \infty$$

and now if we divide both sides by infinity this might look like we can just cancel out the infinity on each side, leaving

$$2 = 1$$

which is another disaster. Maybe you can now guess that something terrible will happen if we think too hard about the last idea:

If you multiply infinity by infinity it's still infinity.

If we write this out we get

$$\infty \times \infty = \infty$$

and if we divide both sides by infinity, canceling out one infinity on each side, we get

$$\infty = 1$$

which is possibly the worst, most wrong outcome of them all. Infinity is supposed to be the biggest thing there is; it is definitely not supposed to be equal to something as small as 1.

What has gone wrong? The problem is that we have manipulated equations as if infinity were an ordinary number, without knowing if it is or not. One of the first things we're going to see in this book is what infinity *isn't*, and it definitely isn't an ordinary number. We are gradually going to work our way toward finding what type of "thing" it makes sense for infinity to be. This is a journey that took mathematicians thousands of years, involving some of the most important developments of mathematics: set theory and calculus, just for starters.

The moral of that story is that although the idea of infinity is quite easy to come up with, we have to be rather careful what we do with it, because weird things start happening. And that was just the beginning of the weird things that can happen. We're going to look at all sorts of weird things that happen with infinity, with infinite collections of things, hotels with infinite rooms, infinite pairs of socks, infinite paths, infinite cookies. Some weird things are like $1 = 0$, not just weird, but undesirable. So we try to build our mathematical ideas to avoid those. But other weird things don't contradict logic, they just contradict normal life. Those weird things don't cause problems to our logic, they just cause problems to our imagination. But it can be very exhilarating to stretch our imagination just like fiction writers do when they create a person who lives infinitely long (immortality) or who can travel infinitely fast (teleportation). And it's not just exhilarating: it can shed new light on our normal life. When characters are immortal in fiction, they often end up realizing how the finiteness of life is actually what gives it meaning.

Infinite Hotels

When we start teaching children the idea of numbers, we usually give them some objects to think about, or we talk about numbers while they're eating something that comes in distinct pieces, like strawberries or beans. Or maybe we count spoonfuls of something as they eat it.

If we try to count spoonfuls of something until we get to infinity, we'll be here for a rather long time. Actually later on we *are* going to do something a lot like counting up to infinity, but for now instead of counting all the way there, we'll play with something that is already infinite: an infinite hotel.

Imagine a hotel with an infinite number of rooms, with room numbers $1, 2, 3, 4, \ldots$ going on forever.

1	2	3	4	5	6	7	8	9	10	.....

Now imagine that you're the manager of this amazing hotel, and every room is full. You're reveling in the amount of money you're raking in when another person arrives and asks for a room. On the one hand, the hotel is full. On the other hand, if you just ask everyone to move up a room . . .

This hotel with an infinite number of rooms is known as *Hilbert's Hotel.* It's named after the mathematician David Hilbert, who used it as a vivid way of illustrating the strange things that can happen when you start thinking about infinity. A normal hotel only has a finite number of rooms. Once it's full that's it – if another guest turns up, there's nothing you can do about it without building an extension. However, with the infinite hotel, you can get the person in room 1 to move to room 2, and the person in room 2 can move to room 3, and the person in room 3 can move to room 4, and so on. We can tell the person in "room n" to move to "room $n+1$," and because we have an infinite number of rooms, every n does have an $n+1$, so every hotel guest has a new room to move into. This leaves room number 1 empty, and the new guest can be accommodated.

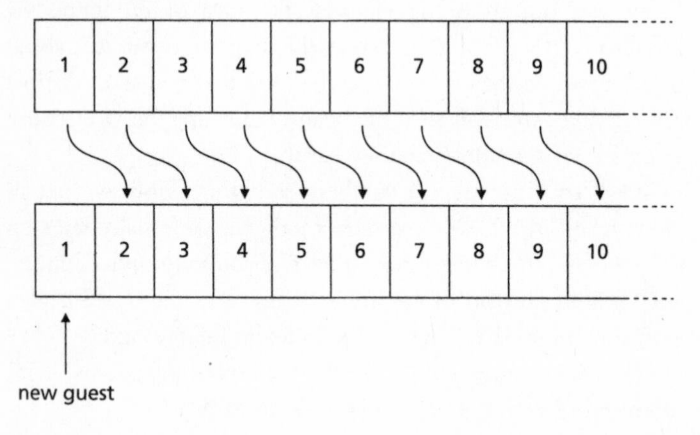

This is sometimes called a paradox, but there's nothing wrong with the argument. It's just that the conclusion is counterintuitive. How can we fit an extra guest into a hotel that's already full? The only reason this is counterintuitive is that we're too used to *finite* hotels. When we start thinking seriously about infinity, as opposed to just vaguely dreaming about infinity, we have to be prepared for slightly strange things to happen. Or even very strange things. This is the fun of infinity.

What we want to do is incorporate "infinity" into normal mathematics without changing the rest of it, just like when people write a book in which one person is immortal but the rest of the world is just as usual. Some strange new things will happen, but we don't want to ruin the basic facts about the world. What that means for infinity and mathematics is that we don't want the presence of infinity to result in $1 = 0$, for example, but maybe some unexpected new things will start to happen, like with this infinite hotel.

Hilbert's Hotel does not cause anything to go *wrong* mathematically – there is no logical contradiction, just a contradiction with our intuition about normal hotels. It widens our eyes to the kinds of things that become possible in the presence of infinity.

What If More Guests Arrive

What if a second extra guest arrives at Hilbert's Hotel? Well, we can just ask everyone to move up a room again. Now the person who was originally in room 1 will be in room 3, and the person who was originally in room 2 will be in room 4, and the person who was originally in "room n" will be in "room $n + 2$." This being the world of mathematics, we don't take into account the hassle of everyone moving rooms, we just feel pleased that everyone has a room.

If those two extra guests had arrived at the same time, we

could have immediately asked everyone to move up two rooms. Of course, if three extra guests arrive, we can ask everyone to move up three rooms. And so on, for any finite number of extra guests.

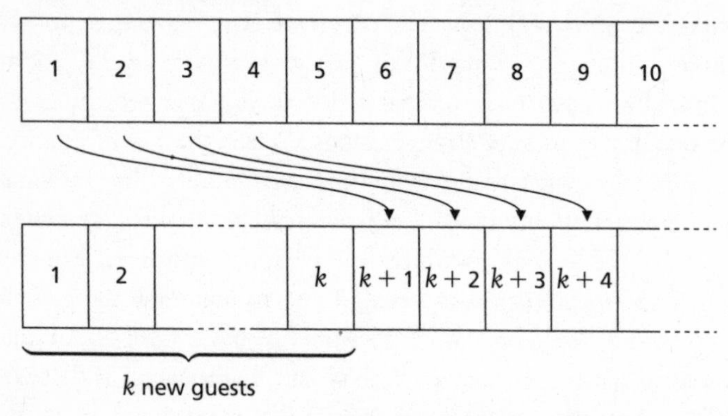

What if an *infinite* number of extra guests arrive? We can't just ask everyone to move up an infinite number of rooms. This might sound like it makes sense, because the rooms go on forever, but let's think about a particular person, say the person in room 1. Which room will they end up in? Room "$1 + \infty$"? That won't work, because it's not a room number. We do have an infinite number of rooms, but each one is still labeled with a finite number. So there is no room "$1 + \infty$" for the person in room 1 to move into. If we can't tell them which room to go to, then we're stuck.

So we have to be a bit cleverer. (Dealing with mathematics usually means having to be increasingly clever all the time, which is why it always seems hard.) We can tell everyone to *double* their room number. So the person in room 1 goes to room 2, and the person in room 2 goes to room 4, and the person in room n goes to room $2n$. This will leave an infinite number of rooms spare. How do we know? We know all the old guests end up in rooms that are double their previous room number, so they all end up in *even*-numbered rooms. This means that all the odd-numbered rooms are now empty, and there are an infinite number of those.

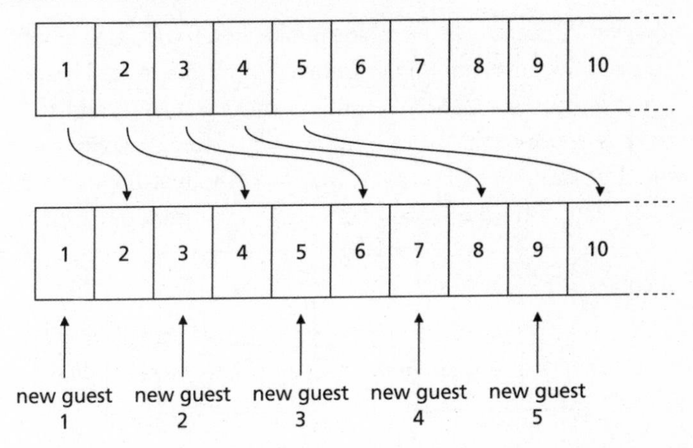

We can tell that this will work for sure by writing down instructions for what everyone's room number is going to be. We could write a very long list, but that would take too long, so instead we can make a formula. The point of a formula in math is to save the effort of writing a long list. Here's how the instructions would go:

* Old guests: if you are in room n, please move to room $2n$.
* New guests: if you are new guest number n, please move to room $2n - 1$.

Now everyone knows which room they're in, and we can check that no two people will get the same room number, unless they did their calculation wrong.

You might notice that this system only works if the new people arrive in an orderly queue, otherwise there will be the mathematical version of an unruly scrum as everyone dives for a room. The new guests all had to arrive in a numbered order so that they could work out their new room number. We're going to focus more on the queueing aspect as things become more complicated.

What If the Hotel Has More Floors?

Now let's imagine we have an infinite hotel with two floors, and each floor has an infinite number of rooms in it. The first floor has rooms $1, 2, 3, 4, \ldots$ and so on, and the second floor also has rooms $1, 2, 3, 4, \ldots$ and so on. (More sensibly these would be numbered $11, 12, 13, 14, \ldots$ on the first floor and 21, 22, 23, 24, ... on the second floor, but never mind for now.)

2nd floor	1	2	3	4	5	6	7	8	9	10	...
1st floor	1	2	3	4	5	6	7	8	9	10	...

What if this hotel has a fire and we need to evacuate all its guests into a one-floor Hilbert Hotel across the road that happens to be empty? This is no problem. We can ask everyone on the first floor to double their room number and subtract one, so they'll go into rooms $1, 3, 5, 7, \ldots,$ just like the evacuated people last time. Then we can ask everyone on the second floor to double their room number, as if they were the original people in the hotel in the last example. So they'll move into room $2, 4, 6, 8, \ldots,$ and so on.

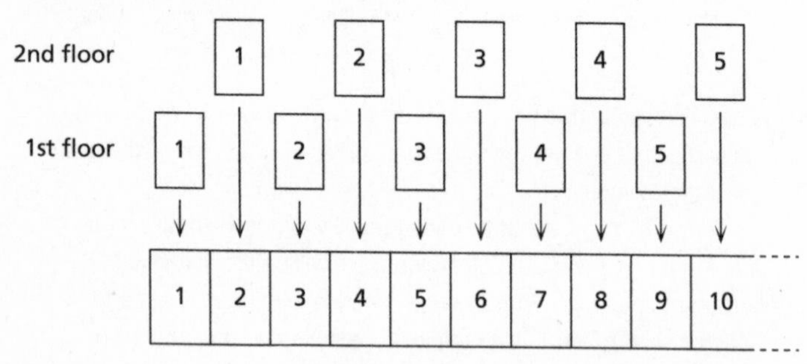

We have, in a way, fit "infinity times 2" people into "infinity" rooms. Mathematically, it's the same problem as fitting infinite new guests into a full one-floor hotel.

This general method will also work with a fire in a three-floor Hilbert Hotel, except this time we have to evacuate "infinity times 3" people, so we have to work on the basis of multiplying everyone's room number by 3.

* The people on the first floor are told to multiply their room number by 3 and subtract 2. So they will move into rooms $1, 4, 7, 10, \ldots$.
* The people on the second floor multiply their room number by 3 and subtract 1. They will move into rooms $2, 5, 8, 11, \ldots$.
* The people on the third floor simply multiply their room number by 3, and they will move into rooms $3, 6, 9, 12, \ldots$.

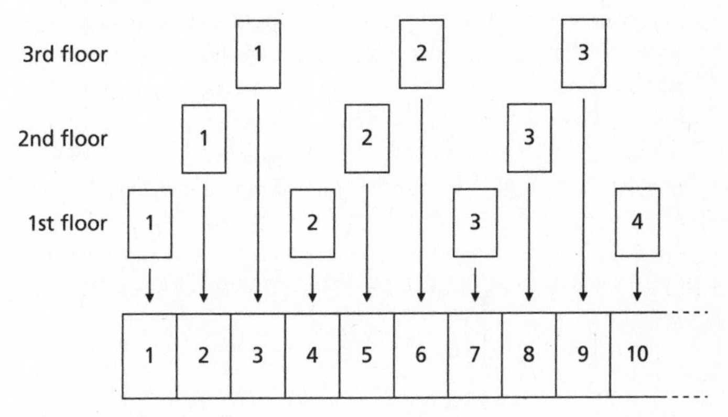

You can also imagine them all queuing up in three queues according to their floor. You would then assign them rooms in order, taking one person from the front of each queue in turn. Whereas if you took everyone from the first queue first and assigned them rooms in order, without having the foresight to leave gaps, you'd run out of rooms.

We can be sure about all this by writing down some instructions again. The right way to write them is like this:

* 1st floor people: if your old room number was n, please move into room $3n - 2$.
* 2nd floor people: if your old room number was n, please move into room $3n - 1$.
* 3rd floor people: if your old room number was n, please move into room $3n$.

Whereas if we took all the people from the 1st floor and filled up the rooms with them in order, we would be saying:

* 1st floor people: if your old room number was n, please move into room n.

But now, are there any empty rooms left for the other people? Impossible, because every room number n is taken up by the person from room number n from the 1st floor of the old hotel. That's why we have to either rotate between the floors as we assign rooms, or assign the rooms to the 1st floor people, leaving gaps between for the 2nd and 3rd floor people, rather than assigning all the rooms in order.

I hope you can imagine this working for any finite number of floors.

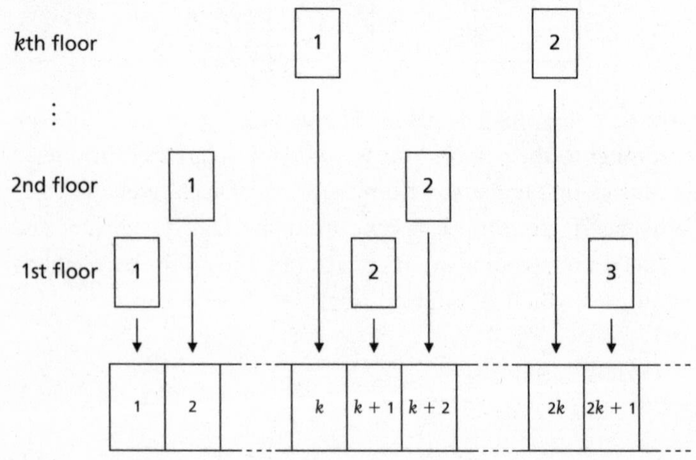

But what about an *infinite* number of floors – an infinite skyscraper Hilbert Hotel, with floors $1, 2, 3, 4, \ldots$ and rooms $1, 2, 3, 4, \ldots$ on each floor. We can think of this as being "infinity times infinity."

⋮	⋮	⋮	⋮	⋮	⋮	⋮	⋮	⋮	⋮	⋮	
5th floor	1	2	3	4	5	6	7	8	9	10	...
4th floor	1	2	3	4	5	6	7	8	9	10	...
3rd floor	1	2	3	4	5	6	7	8	9	10	...
2nd floor	1	2	3	4	5	6	7	8	9	10	...
1st floor	1	2	3	4	5	6	7	8	9	10	...

If this skyscraper has a fire, will we finally be in a hopeless situation? Can we evacuate everyone into a normal, one-floor bungalow Hilbert Hotel? (Perhaps, by now, the one-floor Hilbert Hotel seems like a rather mundane concept. This is often a consequence of stretching our brains to think about more and more extraordinary things – the previously extraordinary things start seeming quite normal. I take this as a sign that we have become cleverer.)

You might think it's hopeless, because we can't just ask everyone to "multiply their room number by infinity" and then subtract something. We can't "rotate through the queues" because this would happen:

- Person 1 from floor 1 would go to room 1
- Person 1 from floor 2 would go to room 2
- Person 1 from floor 3 would go to room 3
- ⋮

- Person 1 from floor n would go to room n
- ⋮

and no rooms would be left for any other people, because every room n would already be taken up with Person 1 from floor n.

However, it's not hopeless – we just have to be yet more clever. The trick is to imagine lining everyone up in queues again, but this time look at the situation diagonally.

these 5th										
these 4th	1	2	3	4	5	6	7	8	9	10
these 3rd	1	2	3	4	5	6	7	8	9	10
these 2nd	1	2	3	4	5	6	7	8	9	10
these 1st	1	2	3	4	5	6	7	8	9	10
	1	2	3	4	5	6	7	8	9	10

If we start in the corner and work our way along diagonals, we can now fit everyone in. This is a bit hard to sum up in a neat formula like we did with the smaller hotels, and is best captured in pictures:

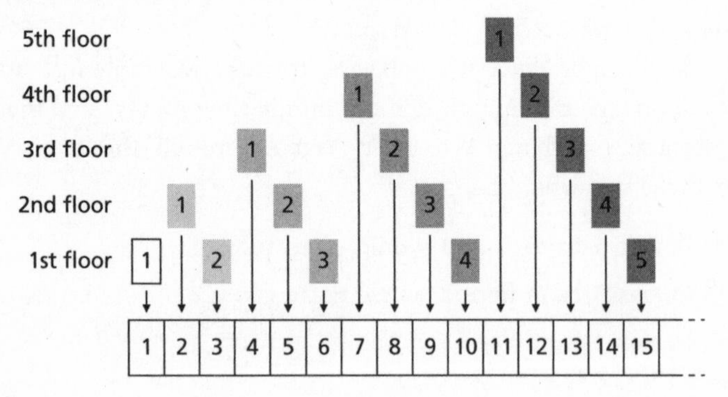

A neat formula for instructing everyone where to go would need to look something like this: Person n from floor k needs to move to room number.... You might be able to work out a formula from the picture, but I think in this case the picture is much more illuminating than the formula.

Incidentally, our discussion about Hilbert's Hotel has actually encoded the following strange fact: there are just as many even numbers as there are whole numbers. Because if you fill up the hotel and then ask everyone to double their room number, you can accommodate exactly the same guests but using only the even-numbered rooms. The fact that we can then fit a whole new floor's worth of people into the odd-numbered rooms means that there are also just as many odd numbers as there are whole numbers. This gives us a hint of how philanthropic we could be if we had an infinite amount of money – we could give an infinite amount of it to charity and still have an infinite amount left. All we'd have to do is give away every even-numbered dollar in our bank account and keep every odd-numbered dollar. This doesn't make sense, because dollars in a bank account aren't usually numbered – you usually just have a total. So instead you could transfer one dollar into a charity account, and one dollar into a separate account of your own, and then one into charity, and one into your own. This would be a bit slow, so instead you could transfer *a billion dollars* into the charity account, and *a billion dollars* into your own account, and keep going like that. You would, of course, have to keep going forever, whatever that means.

Buoyed up by this success, you might feel invincible and think that you'll be able to evacuate *any* hotel into a Hilbert bungalow. However, this isn't true. If we have an even crazier hotel that has a room for every rational *and* irrational number ("Hello, I'm in room number π."), then we will be defeated.

This is all really about the concept of countability, which we're building up to. In Chapter 6 we will see the mind-boggling fact that some infinities are bigger than others. Some things just won't fit into Hilbert's Hotel no matter what you do.

One of the things that is tantalizing about infinity is that it's so easy to stumble upon the idea, so easy to stumble upon strange, apparently magical behavior that surrounds it, yet so difficult to work out what on earth is really going on. We now know that a hotel with an infinite number of rooms is very different from a hotel with a "normal" number of rooms. We know that we can't manipulate infinity in equations like we can manipulate "normal" numbers. It seems like infinity can't be a "normal" number, but what does that mean? Numbers seem to be the most basic building blocks of mathematics, but what are they? This is one of the most basic examples of something we take very much for granted without really knowing what it is. If we're going to claim that infinity isn't a number, we'd better think about what numbers are. You might be surprised to find how long it took for mathematicians to be really sure what numbers are. You might also think it's a bit pointless, seeing as humans got by using numbers for thousands of years without understanding that, and perhaps so have you. Are mathematicians completely irrelevant?

The point is this. Ordinary whole numbers aren't so bad to understand. Even if you extend them to the negative numbers and fractions, they aren't so bad. It's when you get to the numbers between the fractions, the irrational numbers, that things get really tricky. Not knowing what whole numbers really are didn't cause too many problems, but not understanding what irrational numbers are did. Unblocking that obstruction led to the massive mathematical development of calculus that in turn led to huge leaps forward in the precision and understanding of science, medicine, and engineering in the last two centuries. But in order to understand those irrational numbers better, we have to understand all the numbers better, including the most basic ones. We have to get the foundations right in order to build a good, sturdy building, and if the foundations turn out to be shaky, there's nothing left to do but to go back to the beginning and sort them out.

3 What Infinity Is Not

When babies first learn to climb up a step, they are enthralled. They do it again, and again, and again, and they get higher and higher and higher, possibly all the way to the next floor unless some mean adult removes them, which they usually do. I like to think that they have worked out how to construct the whole numbers, by just repeatedly adding one. But will they ever get to infinity like that?

This is part of the big question of whether infinity is a number or not. It's a bit like the infamous lawsuit about whether Jaffa Cakes are cakes or cookies (because for some obscure reason cakes and cookies are taxed differently in the UK). You can't start trying to answer this question without sitting down to decide what the definitions of "cake" and "cookie" are. Are Jaffa Cakes cookies because of their size and shape (small and flat), or are they cakes because of their texture (fluffy and airy) and their behavior (going hard when stale, unlike cookies which go soft when stale)? The lawsuit ruled in favor of texture and behavior, resulting in a "cake" classification and lower taxation.

What about infinity? If we imagine a lawsuit about whether infinity is a number or not, what would be the defining characteristics of numbers that we would come up with? That is what we're going to start thinking about in this chapter. In fact, there are many different types of numbers, starting from the simplest – 1, 2, 3, and so on – moving on to negative numbers, fractions, irrational numbers, and more weird and wonderful things. We are gradually going to rule out the possibility of infinity being any of these.

You might wonder why we can't just declare that infinity is a number and be done with it. To understand this, we have to understand how mathematics works. This might feel a bit like when you look a word up and the definition includes another word you don't understand, and when you look that one up there's another word you don't understand. To understand infinity, we need to understand numbers. To understand numbers, we need to understand mathematics. And, wait for it, to understand mathematics, we need to understand logic.

Mathematics studies things using logic, and so it can only study things that obey the rules of logic. When we say what mathematical objects are, there are various approaches we can take. We can say what they are by *listing* them, or we can say what they are by *characterizing* them. We can say what birds are by writing a great long list of all the birds in the world, or we can say, "Birds are animals with feathers, wings, and beaks." Not only is the second way quicker, it also means you can leave open the possibility of birds you haven't discovered yet. And then, if you find a completely new creature, you have a way of deciding if it's a bird or not.

Here's a mathematical example. We can say, "The platonic solids are the tetrahedron, cube, octahedron, dodecahedron, and icosahedron." Or we can say, "The platonic solids are the convex solids where all faces, edges, and angles are the same." Or, more precisely, "regular convex polyhedra with congruent faces of regular polygons and the same number of faces meeting at each vertex." In the first case we have simply listed them, and it's actually possible, as there are not very many of them. But in the second case we have described what properties they have in a way that enables us to look at an object and decide if it is a platonic solid or not, without having to remember what an icosahedron is.

For some mathematical objects, listing them is too difficult because there are too many of them, like with the birds. For example, we can't list all the numbers in the world because there are infinitely many. Nor can we list all the prime numbers in the world, but for a different reason – we don't know what they are. (But if we did, we still wouldn't be able to list them as there are infinitely many.) In the case of prime numbers we can characterize them as "any number divisible only by 1 and itself (and 1 doesn't count as prime)" and then the challenge is to build up a list of all the numbers that fulfill these characteristics.

Then if we are trying to show that something is *not* a particular type of mathematical object, there are also two ways of doing it. We can either stare at the list and see that it's not on the list, or we can look at the characteristics and show that it doesn't have those characteristics. For example, we can easily see that a sphere is not a platonic solid, simply because it's not on the list of platonic solids. But we can't test if a number is prime by seeing if it's on the list of prime numbers, because that list doesn't exist. Instead we have to see if it behaves in the way prime numbers do: we have to see if it's only divisible by 1 and itself. For example, 6 is also divisible by 2, so we know it can't be prime. If we find a new creature in the forest and want to decide if it should count as a bird or not, we have to decide if it fits the characteristics of a bird. At some point biologists had to decide what the defining characteristics of birds should be, and mathematicians had to decide what the defining characteristics of numbers should be.

There are lists of prime numbers up to a certain size, so as long as your number isn't too big you can in fact check on a list to see if it's prime. However, it has become more efficient for computers to test to see if a number is prime rather than store huge lists of primes on the off chance that someone wants to look one up. The largest known prime number at time of writing has more than 22 million digits. This one number is so long that it's pretty impossible to store it even by itself, let alone to store all the prime numbers up to it. So nobody is keeping a list of all known prime numbers.

Like birds, there are many different types of numbers. Some are more ubiquitous than others, and some are more well known than others. We'll also see that the ones that are the most numerous are not the ones we see or think of the most. We'll start with the most obvious kind of numbers, the ones we probably do think of the most: the ones we count with.

Natural Numbers

The most basic kind of numbers are also the first kind that small children learn to use by counting things: $1, 2, 3, 4$, and so on. In mathematics these are called the *natural numbers*, because they're the most natural. The way I just described them seemed quite obvious – we just count $1, 2, 3, 4$ and keep going. But if we had never learned to count before, how would we know how to keep going? How do we know that the next number is 5 and not $4\frac{1}{2}$? How do we know that $4\frac{1}{2}$ is not one of these numbers?

This is actually a rather profound question. The reason we know that 5 is next is just because we declare it to be so. But what *is* 5? How can we characterize the natural numbers other than by learning all their names?

The way mathematicians do it is basically by saying they're what we get if we keep counting – we just have to make it more precise. We have to make it more precise because we have to make sure it obeys the rules of logic. "What we get if we keep counting" doesn't obey the rules of logic because we don't even really know what "keep counting" means. To make this into something mathematical we have to make it less ambiguous.

So instead of saying "counting" we say "adding 1." The natural numbers are what we get if we keep adding 1 to things. But where do we start? We have to start somewhere, otherwise we'll never get anywhere. And we start at 1.

The natural numbers are what we get if we start at 1 and keep adding 1. In fact there is some argument about whether 0 counts as a natural number, but it doesn't make much difference – we could just as well start at 0 and keep adding 1.

Some people get very excited about whether or not 0 is a natural number. However I think this is just a question of words and names. Mathematically you can start anywhere you want and keep adding 1. Some mathematicians think that 0 is a good place to start because 0 is such a useful number. I can't really imagine anyone disagreeing about 0 being useful, but people really do disagree about whether 0 should be called one of the "natural numbers" or not. I would say this is a question of terminology rather than of mathematics, but even that statement is contentious – some people think that the questions of terminology are part of mathematics. Personally I can just see that arguments about whether or not 0 counts as a natural number can easily put people off the entire subject of mathematics altogether. But some people think it's so important that I will probably receive hate mail for suggesting that it isn't; I have already had people shout at me about it at the end of talks.

If we decide that the natural numbers are "what we get if we start at 1 and keep adding 1," then really they are these things:

$$\begin{array}{l} 1 \\ 1+1 \\ 1+1+1 \\ 1+1+1+1 \\ 1+1+1+1+1 \\ \vdots \end{array}$$

It would be rather tedious to refer to them in this way all the time, which is why we give them new names, a sort of shorthand so that instead of saying "one plus one plus one plus one," we can say "four," which is much quicker. Mathematical language is just there to give us handy ways to refer to long-winded things, but it can seem like a bafflingly large amount of jargon. You might not think of "one, two, three, four" as baffling jargon, but that's because jargon becomes familiar if you use it long enough. If your job is anything like mine, when you first arrived you were dizzy with acronyms people were throwing around, and then a few months later you were throwing them around with everyone else.

When you start a new job someone might give you a handy list of all the acronyms you need to get used to. Alas, we can't list shorthand names for every single possible number, but we can develop a *principle* for building up new names from old ones. You might be able to remember this from learning a foreign language. You usually have to learn the numbers one to ten by rote, and then somewhere between ten and twenty a logical pattern starts to develop. In English this happens after fifteen, as we get six*teen*, seven*teen*, eight*teen* (aside from the extra *t* there), nine*teen*. (You could argue that thirteen, fourteen, and fifteen all fit this pattern but with strange spelling.) In Spanish you also get a pattern from sixteen onward, but in French you have to wait until seventeen. In German you only have to wait until thirteen, and in Cantonese it happens right at eleven, as the numbers are then said as "ten-one, ten-two, ten-three, ten-four, . . ."

Once you've made it to twenty, you usually just have to learn the words for 20, 30, 40, 50, 60, 70, 80, 90, and 100. These follow a pattern with more and less strictness in different languages. In English they vaguely make sense but have some strange spellings. In German they have a pretty good pattern after 20, and in French there's that phenomenon that confuses foreigners, where from 70 upward you count "sixty-ten, sixty-eleven, sixty-twelve" (except in Swiss French where there is a bona fide word for seventy, *septante*) and likewise for 90 upward.

Cantonese is, again, more straightforward, as 20, 30, 40, and so on are simply "two-ten, three-ten, four-ten." I learned from Hindi lecturer Jason Grunebaum that in Hindi there are actually completely separate words for all the numbers from 1 up to 100. He offers bonus credit to any student who can recite them all.

After a hundred, you usually then just have to learn the words for various orders of magnitude: a thousand, a million, a billion, a trillion. (Although in Cantonese there's a specific word for ten thousand, but not for a million, so to say a million you have to say, "a hundred ten thousands," which makes my head spin.)

After that we start running out of words, but we also start running out of uses for them. How often do you need to refer to specific numbers bigger than a hundred thousand million billion trillion? I've certainly never needed to, except when engaging in exuberant hyperbole and referring to American university costing "a zillion dollars" or perhaps "a gabillion dollars."

And yet those numbers still exist, although we have no words for them, just as animals exist whether or not they have been named by humans. Perhaps a better analogy is the planet Neptune, whose existence was deduced using mathematical arguments before it was actually found and named. We know that numbers bigger than a trillion exist just because we can keep adding one forever. In fact we can do a little proof that there must be bigger numbers than we have names for:

> Suppose that such-and-such is the biggest number we have a name for. But we can always add one to such-and-such, which will make a bigger number.

This reminds me of a very sweet scene in the film *Être et avoir*, which follows the extraordinary teacher Georges Lopez and his French village school consisting of a single class of boys from the ages of four to ten. At one point a little boy does something a bit naughty and gets ink all over his hands. Lopez takes the little boy to wash his hands but instead of scolding him starts asking him about numbers. He keeps asking the boy what he thinks is the biggest number, and the little boy keeps answering definitively: he is sure he knows what the biggest number is. First he thinks a hundred is certainly the biggest number. The teacher then gently asks him about a hundred and one. This continues until you see the small boy's eyes widen with astonishment as it dawns on him that this conversation could go on forever.

(It is unfortunate that the film's enormous success resulted in a lawsuit, as it seemed that the filmmakers had made a large amount of money from Mr. Lopez's extraordinary teaching. Public opinion was divided about how much of this money the teacher deserved. The courts decided in favor of: not very much. Some people accused Mr. Lopez of being greedy instead of being the selfless teacher he was supposed to be. It makes me sad that a dedicated and life-changing teacher is not considered to be worth a vast sum of money.)

Infinity Is Not a Natural Number

So far we've seen that we can make all the natural numbers by starting with 1 and adding 1 repeatedly. But how do we know we won't get to infinity like this? In order to answer this we need a slightly more precise way of saying what all the natural numbers are.

The natural numbers are all the numbers that are either

- 1, or
- $n + 1$, where n is itself a natural number.

So 2 is a natural number, because it's $1 + 1$. Also, 3 is a natural number, because it's $2 + 1$, and 2 is a natural number. To show that 10 is a natural number we have to do this 9 times, which is a bit tedious but we will get there eventually.

Can we show that ∞ is a natural number like this? Well ∞ definitely isn't 1, so that rules out the first clause. What about the second clause – is there a natural number n such that

$\infty = n + 1$? Well the trouble is this means that n would have to be $\infty - 1$, but that's still ∞. So ∞ is a natural number if ∞ is a natural number. This is a circular argument, so no help to us whatsoever.

Despite all this work, we still haven't really shown that infinity is not a natural number; we have only shown that infinity is not *obviously* a natural number. But let's pause for a moment and think about the impossibility of what we're trying to do. We don't know what infinity is. If we don't know what it is, we can hardly show that it isn't something in particular.

Or can we?

We don't know exactly how to define infinity, but we do know some things it should do once we've defined it.

* Adding 1 should not make it any bigger.
* Adding it to itself should not make it any bigger.
* Multiplying it by a natural number should not make it any bigger.

We're now going to show that there can't be a natural number that behaves in the way that we would like infinity to behave.

I often say that mathematical objects exist once you've thought of them, as long as they don't cause a contradiction. Here we've "imagined" something called infinity, and we've "imagined" that it behaves in these ways that seem to make sense to us. To see whether that exists we have to show that it doesn't cause a contradiction. Unfortunately, it *does* cause a contradiction! We are going to do a proof by contradiction that infinity can't be a natural number. That is, we assume that it *is* a natural number and show that something goes horribly wrong.

Now, we know certain things are true about natural numbers, such as:

* It doesn't matter what order you add them up, because you always get the same answer. For example, $3 + 2 = 2 + 3$.

* You can subtract natural numbers from each other (as long as you're careful about not going negative, as we haven't introduced negative numbers yet).
* If you do the same thing to both sides of an equation, the equation still holds.

Now let's try applying these to infinity, using our first desirable property of infinity. We start with

$$1 + \infty = \infty.$$

Now we subtract ∞ from both sides, which gives

$$1 = 0.$$

This is simply not true. We only applied one rule here, the idea that we can subtract from both sides, and we produced a falsehood. The logical conclusion is that *subtracting infinity from both sides of an equation cannot be valid*. Since subtracting a natural number from both sides of an equation is always valid, we conclude that infinity cannot be a natural number.

Does this feel circular?

Imagine playing with a small child and declaring affectionately, "You're my favorite little bunny rabbit!" They might well reply, "I'm not a bunny rabbit!" And, in the manner of most adults who are prone to having cute but silly conversations with children, you might insist, "Yes you are!," at which point they might protest, "But I don't have a fluffy tail!"

What they've actually done is a little proof by contradiction:

Suppose I were a bunny rabbit.
Then I'd have a fluffy tail.
But I don't have a fluffy tail.
Therefore I am not a bunny rabbit.

Likewise we have proved that infinity is not a natural number.

> Suppose infinity were a natural number.
> Then we'd be able to subtract it from both sides of the equation.
> But we can't subtract it from both sides of the equation.
> Therefore infinity is not a natural number.

This doesn't mean that infinity isn't a number; it just means that infinity isn't a *natural number*. I was recently called in to arbitrate a dispute between my four-year-old nephew and his best friend. His best friend said, "Infinity isn't a number because my daddy says so and he's a scientist and he knows everything." My nephew, sensibly, focused on disproving the more easily refutable claim about the best friend's daddy knowing everything, and I focused on convincing my nephew that mathematicians know some things that scientists don't . . .

4 Infinity Slips Away from Us Again

"Are we there yet?" is the typical refrain of small children on long journeys. They experience time differently, so ten minutes can already seem like an awfully long journey, which can be a bit trying for the adults if the journey is going to take several hours.

I recently swam back along the coast of Lake Michigan from a walk up a sand dune. The coastline curved out of my line of sight, so I repeatedly thought I must have nearly made it back to our beach of departure but when I arrived it just curved round to reveal more beach. I ended up singing songs in my head to try and get a sense of how long I had been swimming.

Mathematics can sometimes seem like a process of never getting anywhere, because every time you work out something new, it just reveals all the other things you don't know. And it's hard to keep track of where we've come from, because once we understand something, it can be hard to remember how difficult it used to be. I often feel like I'm making no progress in math because everything I already know seems easy and everything I haven't done yet is difficult (otherwise I'd already have done it).

We're going to keep going with our journey through the numbers, now that we know we haven't yet arrived at infinity with the natural numbers. Children accomplish learning how to count, and then almost immediately they have to learn how to "uncount," that is, subtract. As children learn more and more about numbers, they gradually learn about more and more different types of numbers, somewhat following the historical development of numbers but with thousands of years of gradual mathematical discovery compressed into a few years of school. Such is the power of education.

However, we need to cross-examine these types of numbers more closely in order to see for sure whether infinity is one or not. Like many things, it's a level of understanding you don't need until you start asking difficult questions. "What is infinity?" is one of those difficult questions. It can seem futile, dry, or pedantic to spend so long thinking about definitions of things, but I prefer to think of it as revealing. It reveals how the ideas work. If I eat some delicious food in a restaurant, I immediately want to know how it was made. If I am taken on a spectacular journey, I like to see on a map where we went. Investigating what all these numbers really are is the same kind of thing. We are going to see that each new type of number is built from the types you already had, together with a desire for more subtlety or expressive power.

New Numbers from Old Numbers

Mathematicians love building things up from previous things, using as little extra work as possible. This sounds like laziness, but I like to think of it as conservation of brainpower. Our brains are *finite*, our brainpower is finite, and we should save it for the times when we really need it.

There's a joke about mathematicians that goes like this.

> You give a mathematician a saucepan and an egg and ask them to boil the egg. They fill the pan with water and boil the egg. Now you give them a saucepan full of water and ask them to boil the egg. They tip the water out and say, "I have reduced the problem to the previous one."

Building things up from previous things has other advantages besides saving brainpower. It helps us see the relationships between different concepts, and helps us understand how things fit together. Imagine a Baked Alaska recipe that didn't tell you the goal was to put together cake, ice cream, and meringue: it would be rather unclear if you didn't already know what a Baked Alaska was.

We can gradually build up more and more complicated types of numbers starting with the natural numbers. They get progressively less and less natural until they're so unnatural that they get called things like "irrational" and "imaginary." Mathematicians often like to name new concepts with words that come from ordinary life and give us an idea of the characteristics of the new concepts. If we give abstract concepts some character that we can relate to, it can help us feel comfortable with them. I think there's something endearing about an "imaginary number," and a "prime ideal" sounds to me a bit like a juicy cut of beef, but perhaps that's just me. (And prime ideals have nothing to do with what I'm talking about here apart from that they sound tasty.)

The next type of number that we build from the natural numbers is the *integers*. This includes all the natural numbers, and all their negative versions (and zero, if we didn't already have that as a natural number). This comes from our desire to be able to subtract things (if we have such a desire). You might remember being frustrated when you were little and hadn't learned about negative numbers yet. Maybe you were frustrated by the fact that you could add *any* numbers together, at least in principle, but that you were told you could only take away smaller numbers from bigger numbers. (But then I was frustrated about practically everything when I was little.) Well, mathematicians are frustrated by this too. In order to subtract anything from anything else, we need to have negative versions of the natural numbers, and this gives us the integers. The integers are "better" than the natural numbers, in the sense that we can do more things with them.

Mathematics often develops by mathematicians feeling frustrated about being unable to do something in the existing world, so they invent a new world in which they *can* do it. I like to think of us as inveterate rule breakers. As soon as we're presented with a rule saying we're not allowed to so something, we want to see if we can make a world in which we can do it. This is very different from the popular conception of mathematics as a subject in which you have to follow a whole load of rules.

The first thing we demand in the world of integers is zero (if we didn't already have it in the natural numbers). Zero is a special number that has the property that "if you add it to any other number, nothing happens."

$$0 + 1 = 1$$

$$0 + 2 = 2$$

$$0 + 3 = 3$$

$$\vdots$$

$$0 + n = n \quad \text{no matter what } n \text{ is}$$

$$\vdots$$

Next we demand that addition can be reversed or "undone." This is like having a return policy when you buy something. It makes me nervous buying something if I can't change my mind and return it. It doesn't mean I always *do* change my mind and return things. It's just comforting to know that I can.

How do you change your mind and "return" a number once you've added it? You subtract it. Another way to think of it is that you add a negative number on. It's just like when you really do return something in a shop and they give you a receipt with a negative total on it. A negative total will be charged to your card, which in practice means that an amount is subtracted from your bill.

It's important for us to understand subtraction a bit more, because subtraction is exactly what caused us problems with infinity in the last chapter, when we discovered that subtracting infinity from both sides of an equation could not be valid. This is a big clue about what infinity can and can't be, if we can decode it.

Mathematically we say "every number has an additive inverse." This is a number that "undoes" the original number – that is, takes us back to 0. How do you get from 1 back to 0? You subtract 1, or to put it another way, you add -1. How do you get from 2 back to 0? You add -2. How do you get from n back to 0? You add $-n$.

Demanding that every number has an additive inverse is like demanding that all your Lego people have helmets. If you're a spoiled child, then as soon as you make that demand, your parents will acquire Lego helmets for all your Lego people. One of the fun things about mathematics, which is not publicized nearly widely enough, is that you don't have to be a spoiled child to get everything you demand. As soon as you demand additive inverses for numbers – *poof!* – you have additive inverses for numbers. This makes mathematical research great fun: you can have anything that you can think of. The only caveat is that you have to take all the logical consequences of your new toy as well. If you're a spoiled child and ask for a lion, your parents might give you a lion, but the lion might eat you. If you're doing mathematical research and you decide you really want 0 to equal 1, that's perfectly fine, it's just that everything else will have to equal 0 as well, so your world will collapse, as if eaten by a hungry mathematical lion.

If you decide you want 0 to equal 2 or any other number, that's fine; you just end up in a circular world called modular arithmetic. Deciding $0 = 12$ is how we tell the time, and deciding $0 = 360$ is how we measure angles around a compass.

Anyway, as spoiled mathematicians we can decide we want everything to have an additive inverse and – *poof!* – just like that, everything gets an additive inverse. We don't get any extra things, we just get those additive inverses. These are called the integers, and from the way we constructed them we know they must consist of precisely the following things:

* 0,
* any natural number n, and
* $-n$, where n is any natural number.

Written out in a more obvious fashion, these are

$$\ldots \quad -4 \quad -3 \quad -2 \quad -1 \quad 0 \quad 1 \quad 2 \quad 3 \quad 4 \quad \ldots$$

Subtracting 4 is then technically defined as adding the number -4. This sounds a bit convoluted, but to mathematicians it's *less* convoluted, because you don't have to define a new operation (subtraction), you just do the old operation (addition) to new numbers.

Is infinity any of these new numbers? It doesn't look like it. But do we know for sure? It turns out that exactly the same proof by contradiction will work as for the natural numbers. This is because the integers must obey the same rule we used for the natural numbers, that subtracting from both sides of an equation is valid. We know this isn't true of infinity, so infinity can't be an integer. The hunt continues.

Technically what we're doing every time we demand a new type of structure is we're making something "freely." When we made the natural numbers, we were demanding the ability to add 1. The technical name for a mathematical world in which we can add things is a commutative monoid. The natural numbers are the *free commutative monoid on one object*, which means we make sure we can add anything we want. The "free" part is that there are no extra rules or constraints, just addition and more addition as many times as we want. When we make the integers, we're demanding the ability to subtract, which is technically making the *free commutative group on one object*. A commutative group is the technical name for any mathematical world in which we can add *and* subtract things.

Fractions

We have tried to get to infinity using whole numbers without success. So we need a different approach. What about instead of climbing up to infinity, dividing something up into infinite pieces? You might remember being told at school, "You can't divide by zero because it would be infinity." So perhaps we can get our hands on infinity by doing just that: dividing something by zero? Perhaps we can define it as a fraction, $\frac{1}{0}$? Unfortunately that doesn't quite work either. Eventually we'll see that there is some sense to it, but this sense is *not* in the form of an equation,

$$\frac{1}{0} = \infty.$$

It's much more subtle than that.

Imagine dividing a cake up between zero people. How much cake will each person get? This question makes no sense, because there aren't any people. You could say, "All the people get ten cakes each!" because there aren't any people, so it's true that all of those zero people get ten cakes each. But all of those zero people also get twenty cakes each. All of those zero people get forty cakes and sixty-three elephants each. Evidently one divided by zero is not exactly a sensible way to define infinity.

In that case why are we told "you can't divide by zero" if the answer isn't infinity? In mathematics, "why can't I . . ." is usually the wrong question to ask. The right question is "why can I . . . ?" The burden of proof means you have to justify everything you do, logically. This is how all of mathematics is built up, by logical justification. If you can't justify something logically, it doesn't count as mathematics. Just because you can't find a reason *not* to do something doesn't mean you have found a reason to do it. (In practice in my life I need both a reason to do something and the absence of a reason not to. I do mathematics because I love it, not because it's useful; however, if it were useless that would be a reason for me not to do it.)

In any case, to understand all this we have to see what dividing means, really. Division is definitely the hardest of the basic operations $+, -, \times, \div$. It's usually the last one to be introduced at school, and is probably introduced by the idea of sharing things out. After that it might be pointed out without much explanation that it's the "opposite" of multiplication. If you start with 3, multiplying by 4 gives 12, and then to get back to 3 again you *divide* by 4.

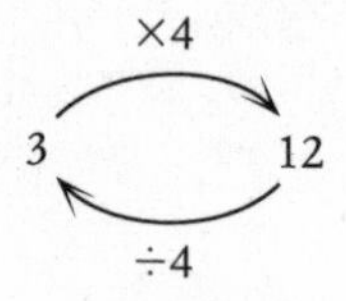

This might remind you of when we "undid" addition to produce negative numbers above. It should, because it's the same process, just undoing multiplication instead of addition.

Just like with addition, we have to start by finding the number that "does nothing" when we multiply it by anything else. This number is 1, because

$$1 \times 2 = 2$$
$$1 \times 3 = 3$$
$$1 \times 4 = 4$$
$$\vdots$$
$$1 \times n = n \quad \text{no matter what } n \text{ is}$$
$$\vdots$$

So 1 is called the "multiplicative identity," just like 0 was called the "additive identity."

We're now ready to think about multiplicative inverses. A multiplicative inverse for a number is the number that "undoes" that number to get us back to 1. What can you multiply 2 by to get back to 1? The answer is $\frac{1}{2}$. What can you multiply 3 by to get back to 1? The answer is $\frac{1}{3}$. What can you multiply n by to get back to 1? The answer is $\frac{1}{n}$.

Now remember how we made the integers from the natural numbers? We demanded, like a spoiled child, that we wanted every natural number to have an additive inverse. Perhaps you now think you want every integer to have a multiplicative inverse. You can do that, but beware: a mathematical lion will appear and eat everything. This is because *demanding a multiplicative inverse for* 0 *is a mistake.*

Here's how things go wrong if you demand a multiplicative inverse for 0. Let's call this number x, because we don't actually know what it is yet. We just know that multiplying 0 by it gets us back to 1, that is

$$0 \times x = 1.$$

But wait. $0 \times x$ is always 0.

This actually takes a little effort to prove. Basically you have to say $0 + 0 = 0$ and so $(0 + 0)x = 0x$ but by the distributive law the left-hand side is $0x + 0x$, so we get $0x + 0x = 0x$. Now subtract $0x$ from both sides. On the left we get $0x$ and on the right we get 0. So $0x = 0$.

So the equation

$$0 \times x = 1$$

becomes

$$0 = 1.$$

Oh dear, that same nonsensical statement has popped up yet again. Are we ever going to escape it? The moral of that story is that we really must not ask for a multiplicative inverse for 0, because if we have one, then we'll also get $0 = 1$ and then *everything* will equal 0.

Finally, remember how we said that subtracting a number is really "adding its additive inverse"? Well, likewise, dividing by a number is really "multiplying by its multiplicative inverse." So dividing by 2 is really multiplying by $\frac{1}{2}$, and dividing by 3 is really multiplying by $\frac{1}{3}$. So can we divide by 0? To do so we would have to multiply by its multiplicative inverse, and we have just decided that 0 is definitely not allowed to have a multiplicative inverse. And because it doesn't have a multiplicative inverse, we have no way of dividing by it. And since we have no way of dividing by 0, we definitely can't define infinity to be $\frac{1}{0}$ because it doesn't exist. Worse than that: it *can't* exist, so we can't just dream it into existence like we did with the negative numbers.

Constructing the Rational Numbers

The technical name for all the fractions is the rational numbers. Rational numbers aren't numbers that sit around having very logical conversations with one another. Rather, they are numbers formed by taking *ratios* of the previous numbers we built up – the integers.

When we constructed the integers from the natural numbers it was quite easy – we just acquired additive inverses for everything, and that was the end of that. Constructing the rational numbers from the integers is a bit more involved. We start by acquiring multiplicative inverses for everything except 0, but this will only give us things like $\frac{1}{2}, \frac{1}{3}, \frac{1}{4}$, and the negative versions, whereas it won't give us things like $\frac{4}{5}$ or anything with a number other than 1 on top. To get those we have to make another spoiled-child-like demand: we want to be able to multiply everything by everything else. We didn't have to do this for addition in the integers, because demanding all the negatives already gave us all possible answers to addition. However, acquiring all possible multiplicative inverses doesn't automatically give us all possible answers to multiplication. If we just throw in the multiplicative inverses, we'll only get $\frac{1}{n}$, where n is any integer. So in total we'll have the numbers:

$$\ldots \quad -4 \quad -3 \quad -2 \quad -1 \quad 0 \quad 1 \quad 2 \quad 3 \quad 4 \quad \ldots$$

$$\ldots \quad -\frac{1}{4} \quad -\frac{1}{3} \quad -\frac{1}{2} \quad -\frac{1}{1} \qquad \frac{1}{1} \quad \frac{1}{2} \quad \frac{1}{3} \quad \frac{1}{4} \quad \ldots$$

If you try multiplying two numbers from the top row, you'll get another number in the top row. If you try multiplying two numbers from the bottom row, you'll get another number from the bottom row. But if you try multiplying a number from the top row with a number from the bottom row, you might get an answer that's in neither row.

So in order to be able to multiply, we have to acquire all the products of top- and bottom-row numbers. This gives us all the rational numbers, which we can summarize like this:

The rational numbers are all the fractions $\frac{a}{b}$ where

* *a and b are integers,*
* *b is not 0, and*
* *$\frac{a}{b} = \frac{c}{d}$ if $ad = bc$.*

The last point is to make sure we don't accidentally think that $\frac{1}{2}$ and $\frac{2}{4}$ are different numbers.

That was quite hard work but hasn't gotten us any closer to infinity. We can use the same argument as before to show that infinity isn't a rational number, because rational numbers also have to obey the rule about subtracting from both sides of the equation being valid. Making the rational numbers has not enabled us to pin down what infinity is. The mathematical coastline continues to curve round out of sight.

Irrational Numbers

Irrational people are people who are not rational, and irrational numbers are numbers that are not rational. You might have been taught that these are "decimals that go on forever" or "the square roots of things." Both of these statements have *some hint of truth* to them, but neither is rigorously true.

Let's take decimals that go on forever. If we try to expand $\frac{1}{9}$ as a decimal, we will get 0.1111111111 . . . and those 1's will in fact go on "forever." But $\frac{1}{9}$ is definitely rational. The reason that the decimal expansion goes on forever is that we've arbitrarily chosen *base* 10 for decimal expansions, and 10 and 9 don't go very well together. Whereas 10 and 5 go very well together, which is why $\frac{1}{5}$ is a neat and tidy decimal 0.2. Here "go well together" actually means "have a common factor."

> If we expanded $\frac{1}{9}$ in base 9 we would get a perfectly tidy 0.1, and if we expanded it in base 3 we would get a tidy 0.01.

The truth behind the "going on forever" statement is that irrational numbers have decimal expansions that go on forever *without ever repeating*. This is quite fun to prove (at the level of, say, an undergraduate analysis class) and satisfying to think about, but a hopeless way of actually checking whether or not something is rational. Even if you could write down a decimal expansion to a million digits (which would take about a week without sleep), how do you know it doesn't start repeating itself after two million digits? Or after a billion or a trillion digits?

The other statement about "square roots of things" is also problematic, because the square root of 4 is 2, which is blatantly rational. And then there are irrational numbers like π and e, which aren't the square root of anything helpful. I know, π is the square root of π^2, but this doesn't help us see that π is irrational. This is why the only good definition of an "irrational number" is "a number that *cannot* be written as a ratio of integers."

You might want to object at the moment and claim that $\pi = \frac{22}{7}$. However, this famous fraction is only an *approximation* to π and was handy in the days before there were calculators, let alone calculators on phones with a π button. π and $\frac{22}{7}$ only agree in the first two decimal places, 3.14. This is perfectly good enough for whenever I need π in my daily life (which mostly involves scaling cake recipes up or down), but it's very far from being mathematically equal. Actually, when I'm scaling a cake recipe I use 3 as a good enough approximation to π.

You might now feel that it must surely be impossible ever to prove that a number *really can't* be written as a ratio of two integers, but there are some fairly short and sneaky proofs of this for numbers like $\sqrt{2}$. The proofs for π and e are much more complicated, it's true.

In any case, none of this helps us say what irrational numbers *are*. We can't just say that they're "everything that can't be written as a ratio of integers," because that would mean that an elephant is an irrational number. We can't say they're "all the numbers that can't be written as a ratio of integers," because what's a number? I've merrily mentioned the number e above, but what is e? Does that mean f, g, and h are numbers? If π is a number, is α a number?

Constructing the irrational numbers is *very hard*, and we'll postpone more of this discussion until later in the book when we're talking about infinitesimally small things. But the idea is just like before: we make a spoiled-child demand for something we want to be able to do, and we acquire all those things that we need in order to do it. The question is, what is it that we want to be able to do? The answer is something like "fill in all the gaps between the rational numbers." Because the fact is that there is always a gap between any two rational numbers. We can depict the integers on a number line,

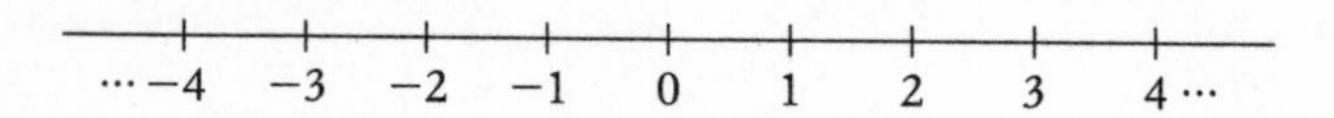

and we know that there are rational numbers between those integers,

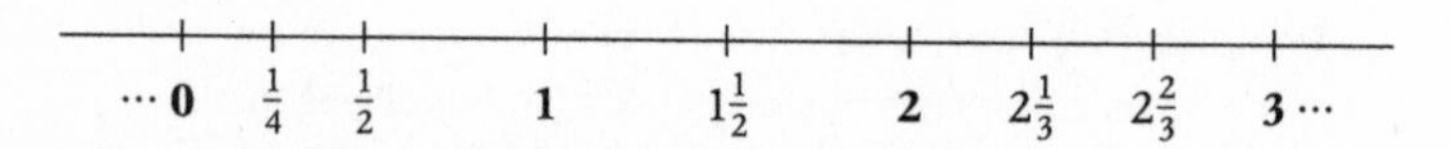

but even after that there will still be some gaps. Without yet being able to explain how we fill them in, here are some amazing things that happen when we do.

* Between any two rational numbers is an irrational number.
* Between any two irrational numbers there is a rational number.

This makes it sound a bit like the odd and even numbers, which alternate neatly along the integer number line. And yet, here's the really counterintuitive part:

There are more irrational numbers than rational numbers.

How can that be possible? If they alternate, how can there be more of one than another? The answer is: this is yet another mysterious thing that happens because we are thinking about infinity. In this case we are not only thinking about an infinite number of things (numbers), but we are also thinking about things being *infinitely close together*. With the odd and even numbers, we can pick any even number, say 6, and know exactly which are the odd numbers right next to it: 5 and 7. However, with rational and irrational numbers, they are so close together that we can never say which are the irrational numbers right next to any given rational numbers. Whichever ones we name, there'll be some more that are closer.

> Suppose you start with the number 1. What is the first rational number that is bigger than it? There isn't such a thing. If you claim it is some number x, I can always make the number $\frac{1+x}{2}$. This is like taking the average of 1 and x, so it will be halfway between 1 and x, therefore bigger than 1 but smaller than x. And it will still be rational because if x is rational then $1 + x$ is rational, and dividing it by 2 still leaves it rational. You could keep splitting the distance and getting closer and closer to 1 with rational numbers that were smaller and smaller but still bigger than 1. Just like dividing the remains of a chocolate cake in two repeatedly to make it last forever, you would never actually reach 1, and there would still be rational numbers left between your latest number and 1.

You can think of this as a process of zooming in repeatedly on the number line, maybe on your computer screen. If you zoom in on the number 6 on an integer number line, eventually you'll be so close up that the only integer that fits on your screen is 6, because the others will have fallen off the sides. Whereas if you zoom in on a number line with all the gaps filled in, no matter how much you zoom, there will still be more rational *and* irrational numbers. So we can't really say they "alternate." It's much weirder than that.

The rational and the irrational numbers together are called the real numbers, and this zooming-in property we've been discussing is called the *density* of the rationals and the irrationals inside the reals. The rational numbers aren't *all* the real numbers, but they're so dense in there that no matter how closely you look, you can't get away from them. Likewise the irrationals. We'll look at this again in Chapter 15 once we understand more about infinitesimally small distances.

Does any of this help us say what infinity is? No. We can still show that, for the real numbers, subtracting things from both sides of an equation is valid, so if infinity were a real number, we would still cause that pesky contradiction $0 = 1$. So infinity is not a real number.

You might be starting to realize now that this hunt for infinity is going to be completely futile unless we somehow come up with a type of number where subtracting from both sides of an equation is *not* valid. What kind of number could that possibly be? So far, every time we have found a new type of number we have done it by demanding the ability to do more things: subtract, divide, fill in gaps. Every time we demand to do more things, we land in a world with more building blocks. It's like if you demand to be able to cook more recipes, you need a kitchen with more ingredients.

Oddly enough, what we really need is a world in which we can do *fewer* things, because we need a world in which subtracting from both sides of an equation is not valid. Sometimes when I'm packing for a trip I have so many different activities planned that I find myself throwing more and more things in my suitcase for all the different events. I don't want to be that type of mathematician who shows up for every event wearing the same clothes, whether it's a lecture, a hike, a concert, or a beach party. However, eventually I'll have thrown in so many different pairs of shoes that I can't lift the suitcase anymore, which means I can't go anywhere, let alone do anything. We have reached this point with our numbers as well. We need to throw everything out and start again, from the beginning.

5 Counting Up to Infinity

I was once walking home as usual when I saw an interesting new clothes shop. I was curious as to why I hadn't seen any building work going on, so I asked them how long they'd been open. They said ten years.

I don't know why I suddenly noticed that shop that day when I hadn't seen it any of the previous thousand or so times I must have walked past it. Sometimes going more slowly or having a slight change in perspective can make us see completely different things in places we thought were familiar. We're now going to change our perspective on numbers slightly and find that this time infinity will jump out at us.

We have tried counting up to infinity and it didn't work. But here's a surprise: it's because we were being too fancy. When children first learn to count they don't do it by adding one repeatedly: they do it by counting on their fingers.

It's just like my favorite way to eat Hula-Hoops, just like most small children, I dare say. American readers might not know these ring-shaped potato snacks, but I'm told on good authority that Americans do this with Bugles, although they're cone-shaped rather than ring-shaped: you put one Hula-Hoop (or Bugle) on each finger, waggle them around a bit, and then eat them off your fingers one by one.

Counting on your fingers is usually thought of as an unsophisticated counting method, something that small children do before they've mastered the art of counting in their heads. I'd like to present a different point of view: counting on your fingers is *very profound*, much more efficient, and gets us to infinity. At least, this idea gets us to infinity, even though our fingers themselves don't.

The brilliant thing about counting on your fingers is that you don't need to keep track of where you are at any given moment, and you don't even have to go in order. You just have to assign one finger to each thing you're counting, and stop when all your fingers have been used up – that is, if you're trying to count ten things. One reason this is efficient, at least for me, is that it saves me the brainpower of having to remember what number I'm on, and frees up that brainpower for doing something else. Instead of remembering that I'm on four, say, I just have to keep the same fingers held down, which doesn't stop my brain doing something else complicated at the same time. You might think I'm being ridiculous – I'm a mathematician after all. Am I just making one of those silly jokes about mathematicians being bad at counting?

Here's the situation where it arises most often. I have a friend round for coffee, and I'm trying to count four scoops of coffee to put in the coffeemaker (I like two scoops per person). I'm also trying to continue a conversation with my friend, in order to be friendly. Typically I will lose count somewhere between two and three – I will pause and have no idea whether I've put two or three scoops in so far. The part of my brain that's holding the conversation seems to have interfered with the part of my brain doing the counting. However, if instead of counting in my head I count with my fingers, I can get it right. (I realize that someone less friendly than me might not mind halting the conversation, and someone less mathematical than me might not mind a more vague quantity of coffee. But I am both mathematical and friendly!)

So much for the efficiency aspect. Here's why I think counting on your fingers is also profound. First of all, try to define ten mathematically. *What is ten?* Earlier on we said it was defined to be:

$$1 + 1 + 1 + 1 + 1 + 1 + 1 + 1 + 1 + 1.$$

But if you didn't previously know the concept of "ten," how could you find ten of anything?

Ten is something that matches up with the number of fingers we have. It also matches up with any other set of things that matches up with the number of fingers we have. This doesn't sound very mathematical does it? Actually when mathematicians finally got round to defining numbers rigorously, this is pretty much how they did it (though not exactly with reference to fingers).

Here is how counting ten things works. Suppose you want to count ten snakes, like in the homework question a friend showed me – her child had been given some "real-life" math questions involving all sorts of "real-life" situations, like counting snakes.

First we decide that ten stands for "the number of fingers we have." Then we make up a name for each finger so that instead of physically using our fingers, we can say their names out loud (or in our heads) to remember how we've matched the snakes up with our fingers. We might call our fingers Tom, Steve, Pieter, Nick, Richard, Emily, Dominic, John, Neil, and Alissa . . . or we might call them one, two, three, four, five, six, seven, eight, nine, and ten. Then we give these names to the snakes, and when there's one snake with each of those names, we know that we've matched up the snakes to our fingers. We can sum up this elaborate process in this diagram

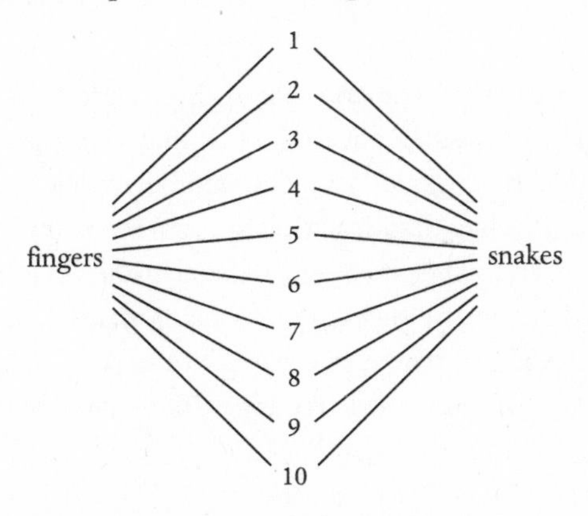

The diagram shows how we've inserted an extra step in the middle. Instead of matching the snakes with our fingers, we've first matched our fingers with names, and then matched the names with snakes.

One reason to do this, of course, is that we often need to count beyond ten, in which case we run out of fingers. You can try using your toes, or you can try using your fingers as ten digits of binary, which means you can use your fingers to count up to 1023. I will do this in Chapter 7.

However, skipping the step in the middle is sneaky, efficient, and useful not just for me and my conversationally impaired coffee-making brain. Suppose you've booked a table for sixteen in a restaurant and all your friends are milling around the table chatting. It can be quite hard to see how many people there are as everyone is moving around; it's much easier to get everyone to sit down and see if there are any chairs left over. You haven't directly counted people, you've just had a look to see if the people match the chairs.

This is the mathematical version of counting. You might think this is an odd statement – isn't counting inherently mathematical already? What I mean here is: this is how mathematics makes the notion of counting *rigorous*.

On Rigor

I'm going to pause now and take a brief digression on mathematical rigor. Mathematical rigor is the thing that enables mathematicians to agree with one another about what is and isn't correct, rather than just having arguments about competing theories and never coming to a conclusion. Mathematics is based on the rules of logic, the idea being that if you only use objects that behave strictly according to the rules of logic, then as long as you only strictly apply the rules of logic, no disagreements can ever arise.

However, if you use objects that don't behave according to logic (like human beings, or clouds), then different valid answers can occur. And if you apply something other than a strict rule of logic, you can also get different outcomes. The world, in general, does not behave according to strict logic. If you give a child a cookie and then another cookie, the chances are they won't have two cookies but zero cookies (unless you count the ones in their stomach).

Mathematics starts with the process of stripping away the ambiguities and leaving only things that can be unambiguously manipulated according to logic. It continues by then manipulating those things according to logic to see what happens. It can be frustrating both in your own life and in the history of mathematics, because things that feel very "obvious" turn out to be very difficult to make unambiguous. So what is the point of doing it?

One point of doing it is to gain access to things that *aren't* obvious. If it's not obvious by our intuition, then we have to find some other way of getting there. One example is infinity. We seem to be having trouble working out how to deal with infinity in a way that makes sense. All our attempts so far have caused us to conclude that $0 = 1$, which won't do at all. So what we're going to do is rethink our approach to *finite* numbers in a way that will enable us to think about infinity as well. We're going to rethink our approach to the numbers we thought were the most obvious and basic: the counting numbers.

Counting by Bags

Counting is, essentially, a process of matching up one set of things with another "official" set of things that defines the number in question. We've already seen that we can define ten by our fingers – our fingers are the official set of things that defines "ten," and whenever another set of things can be lined up with our fingers, that counts as ten things.

We could designate one official set of things for each number. We could, for example, have an official bag of things for "23," and then 23 is defined to be "the number of things in that bag," and everyone who needed to count 23 things would have to match them up with the things in that bag. This sounds very silly and made-up, but compare it with the real definition of a kilogram: a kilogram is technically defined to be the mass of an official lump of metal that sits in a vault near Paris.

Mathematics doesn't really define numbers according to official bags of objects that exist physically, but that's because mathematics is an abstract subject, not a physical one. So it defines numbers according to official "abstract" bags of objects. These bags are abstract because they don't exist physically, but they exist as ideas. And mathematics doesn't call them bags: it calls them sets. But we can think of them as bags.

The first bag we think about is the bag containing zero things: an empty bag. We define zero to be "the number of things in this bag." Perhaps we write a big 0 on this bag.

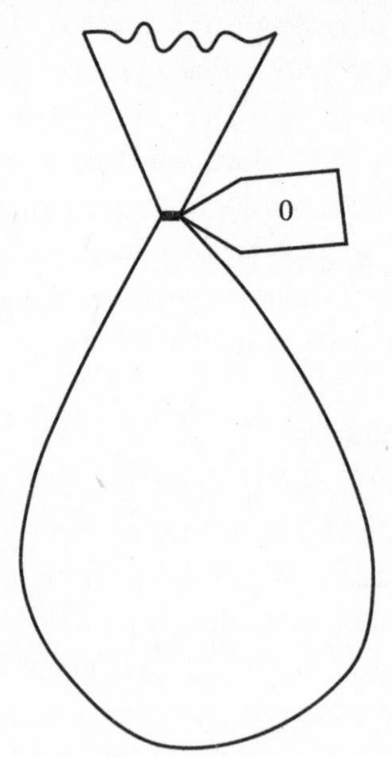

Now we need to make a bag containing one object. Look around you – what objects do you have to work with? Well, so far all we have is an empty bag. And that will do. So we make a bag containing one empty bag. This defines the number one. Perhaps we write a big 1 on this bag.

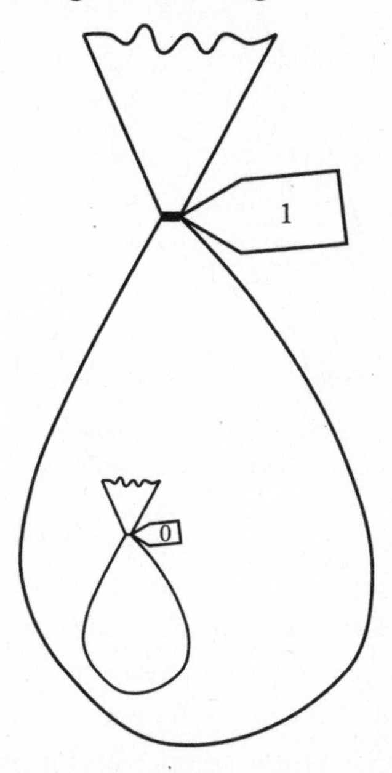

Now we need to make a bag containing two objects. And we do in fact have two objects right in front of us: the "zero bag" and the "one bag." So we make a bag containing these two things and write a big 2 on it.

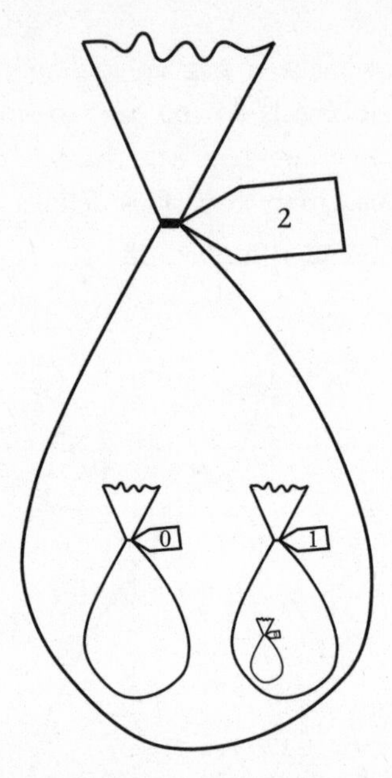

It's very easy to get confused now – you might protest that in total we have three bags inside the bag number 2. But you're not supposed to look inside the inner bags; you're just supposed to take them as objects and ignore the fact that they might contain things. Just like when you open a bag of fun-size bags of M&M's, you can count the number of fun-size bags in the big bag without counting the number of M&M's in each small bag.

Here's bag number 3:

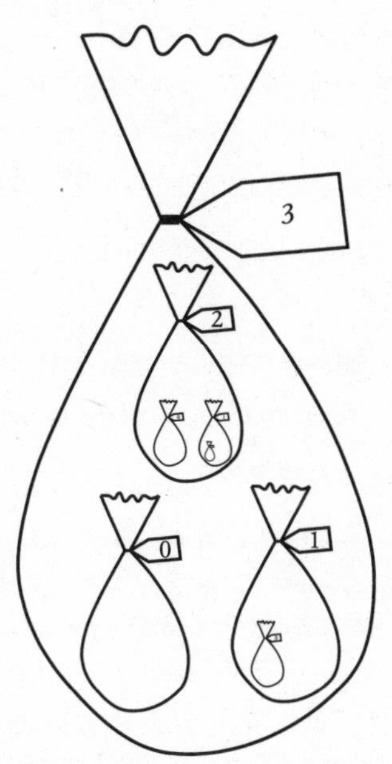

Speaking of M&M's, perhaps you're wondering why we don't fill our bags with more sensible counting objects like pebbles or coins. The answer is that although these objects seem more sensible things to count with, they don't exist in the abstract mathematical world. Bags, or rather sets, are the only starting point that we're given here, so we need to try and build up our whole mathematical world from just this notion of set. Or at least, that's the approach called *set theory*. There are other approaches to the foundations of mathematics, but set theory is the one that came up with this rigorous way of defining infinity.

Once we've got our "official" bags of 0 objects, 1 object, 2 objects, 3 objects, and so on, we can start counting how many things are in other bags by matching up the objects in the new bag with one of the official ones. We have to match up the objects in a careful way:

1. Every object in the new bag has to be paired up with an object in the official bag.
2. You can't have two objects in the new bag being paired up with the same object in the official bag.
3. You have to use up every object in the official bag.

It's a bit like pairing up the professionals with the celebrities on *Dancing with the Stars*:

1. Every celebrity has to be paired up with one pro.
2. You can't have more than one celebrity with the same pro.
3. You have to use every pro.

As a result, you know that there are the same number of celebrities as professionals. Whereas if two celebrities could be paired with the same pro, there might be more celebrities than pros, and if some pros weren't used, then there might be more pros than celebrities. This wouldn't *necessarily* be the case. You could have the same number of celebrities as pros, but two celebrities paired with the same pro and one pro sitting around doing nothing. This is perhaps a bit clearer to see in diagrams.

Laurie Hernandez	⟶	Valentin Chmerkovskiy
Marilu Henner	⟶	Derek Hough
Amber Rose	⟶	Maksim Chmerkovskiy
Ryan Lochte	⟶	Cheryl Burke
Vanilla Ice	⟶	Witney Carson
Maureen McCormick	⟶	Artem Chigvintsev
Calvin Johnson Jr.	⟶	Lindsay Arnold
		Emma Slater

From this diagram we can see that every celebrity is paired with exactly one professional, but there is one professional, Emma Slater, who does not seem to have a partner. We can conclude that there are more professionals than celebrities listed, even without counting them.

However, if several celebrities were all paired with Lindsay Arnold, say, as well as Emma Slater having no partner, then it would be harder to tell if there were more celebrities or pros without counting them.

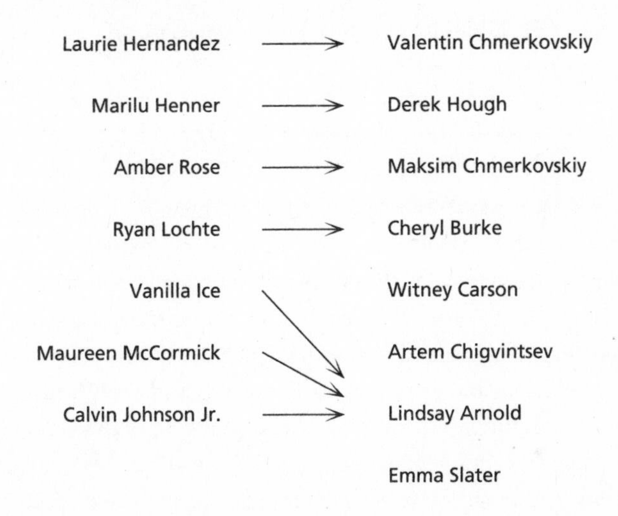

This business of pairing things is called a *function* in mathematics. Suppose you have two sets of objects, let's call them C for celebrity and P for pro. A function from C to P is a pairing up of objects in the set C with objects in the set P, satisfying the first condition above: everything in C has to have exactly one partner in P.

The other conditions are special properties that are only satisfied by special functions. The second condition, saying "you can't have two celebrities with the same pro," is called *injectivity*. The third one, saying "you have to use every pro," is called *surjectivity*. When we have all three conditions, we can think of it as a *perfect pairing*, where everyone on each side has exactly one partner on the other side; nobody has more than one partner and nobody is left out. The picture would look like this:

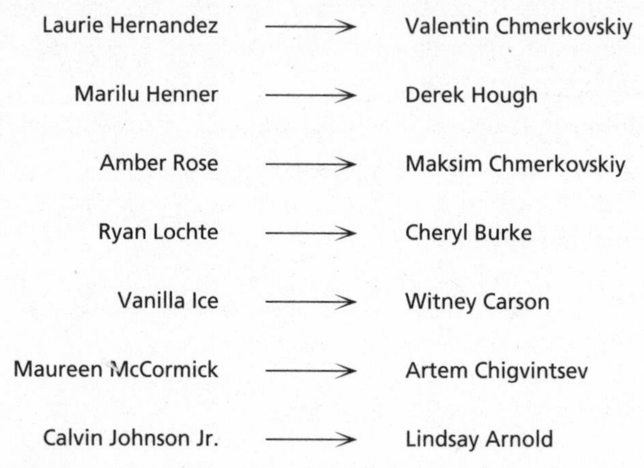

The technical word for this is a *bijective* function, or a bijection.

You might wonder why we don't need extra conditions saying "you can't have two pros with the same celebrity" and "you have to use every celebrity." It's because these are inherent in the first condition, which is the very definition of function. The celebrities and pros play different roles in *DWTS* just like they do with functions. Another way of thinking of this is that it's more like a vending machine. The things on the left are the buttons you push (or codes you punch in) to choose products, and the things on the right are the products you can buy. It is possible that several different codes give you the same product – for example, there are those vending machines where every position is filled with cans of Coke. But it wouldn't make sense if the same code could give you different possible products. Or rather, it would be frustrating, because you couldn't be sure which product you were going to get. That is a possible type of vending machine, but not one we're talking about now.

This idea of pairing things up is going to give us our first valid definition of infinity, so let's see how this works for some more mathematical situations. The technical way of talking about functions is very dry, but we can also draw pictures of the pairings. The pictures help us to get a feel for what is going on, but the dry technical way is much better for mathematicians to do unambiguous work and be sure of not making mistakes. Unfortunately, mathematics is often explained with just the dry part and not the feeling.

* Here's a picture of a pairing much like the *DWTS* example, but with numbers instead of people.

$$1 \longrightarrow 4$$
$$2 \longrightarrow 8$$
$$3 \longrightarrow 12$$
$$16$$

More precisely: C is the set containing the numbers 1, 2, 3.
P is the set containing the numbers 4, 8, 12, 16.
The function takes each number in C and pairs it with the number that is four times it. So 1 is paired with 4, 2 is paired with 8, 3 is paired with 12. But 16 is sitting around doing nothing. This function does not satisfy the third condition; that is, it is not surjective. You can see this because, like Emma Slater in the previous example, 16 does not have a partner.

* Here's an example where two "celebrities" are paired with the same "professional."

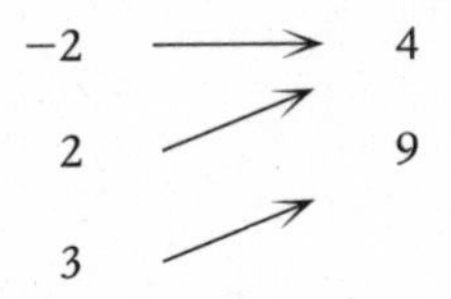

C is the set containing the numbers -2, 2, 3.
P is the set containing 4 and 9.
The function takes the numbers in C and pairs them with their squares. So -2 is paired with 4, 2 is also paired with 4,

and 3 is paired with 9. Here -2 and 2 are both paired with the same object in *P*, so the function does not satisfy the second condition, that is it is not injective. We can see that, as there are two arrows pointing to the same place.

* Here's an example where the "celebrities" and "pros" are perfectly paired up.

$$1 \longrightarrow 5$$

$$2 \longrightarrow 6$$

$$3 \longrightarrow 7$$

C is the set containing the numbers 1, 2, 3.
P is the set containing the numbers 5, 6, 7.
The function takes the numbers in *C* and pairs them with the numbers that are 4 bigger. So 1 is paired with 5, 2 is paired with 6, 3 is paired with 7. We can see that no two arrows are pointing to the same place, and no number on the right is sitting around doing nothing. This is a perfect pairing of numbers in *C* and *P*. It means we know there must be the same number of things in each set, just like when we paired up the snakes with our fingers.

✻ Now let's be a bit daring and try this with some infinite sets.

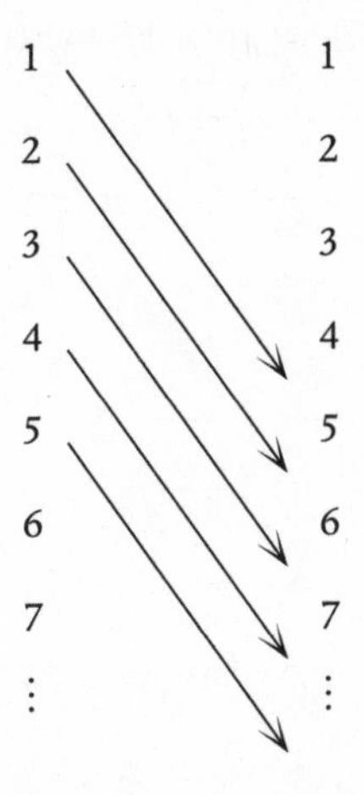

C is now the set of all natural numbers.
P is also the set of all natural numbers.
The function takes the numbers in *C* and pairs them with the numbers in *P* that are four bigger, just like in the last example above. It is still true that no two numbers in *C* are paired with the same number in *P*, but this time several numbers in *P* are left out: the numbers 1, 2, 3, and 4. So this function is injective but not surjective.

* Here's one where we pair up all the natural numbers with just the even numbers, a bit like we did when we had everyone in Hilbert's Hotel double their room number.

$$
\begin{array}{ccc}
1 & \longrightarrow & 2 \\
2 & \longrightarrow & 4 \\
3 & \longrightarrow & 6 \\
4 & \longrightarrow & 8 \\
5 & \longrightarrow & 10 \\
6 & \longrightarrow & 12 \\
7 & \longrightarrow & 14 \\
\vdots & & \vdots
\end{array}
$$

C is the set of all natural numbers.
P is the set of all *even* natural numbers.
The function takes each number in C and doubles it, pairing the original number with the doubled version. No two arrows point to the same place, and also every number in P gets used. So this is another perfect pairing.

* Now for an example that's like evacuating a two-floor Hilbert Hotel into a one-floor hotel. We start with red numbers and blue numbers on the "celebrity" side and purple numbers on the "professional" side.

red	blue		purple
	1	⟶	1
1		⟶	2
	2	⟶	3
2		⟶	4
	3	⟶	5
3		⟶	6
	4	⟶	7
4		⟶	8
⋮			⋮

C is now the set of natural numbers, twice over: all the natural numbers in red, and also all the natural numbers in blue.

P is just one set of natural numbers, in purple.

This function works just like the formula for evacuating the two-floor Hilbert Hotel on page 20: the red number n gets paired with the purple number $2n$, while the blue number n gets paired with the purple number $2n - 1$. We can see from the picture (as well as from our hotel discussions) that this does satisfy both the conditions: no two arrows point to the same purple number, and all the purple numbers do get used. So this function is also a perfect pairing.

These last few pictures showed us how we can make pairings with infinite sets and they could still fail to be perfect pairings. Earlier we said that if you have a perfect pairing – that is, a bijective function between two sets – it shows us that the two sets have the same number of objects in them. If you can pair up celebrities with pros precisely, you know there are the same number of each without having to count them one by one. The clever part is that because you don't have to count them one by one, it still works for infinite sets and is the key to unlocking infinity. It is the mathematical notion of countability.

Countability

Learning to count sounds like the first thing children do. Surely everyone learns to count when they're very small? This is true but it doesn't mean that everything is then easy to count. There are various ways things can be difficult to count in normal life. They could be moving around a lot, like children on a playground or rabbits in a field. They could all look alike so you can't really tell which ones you've already counted, like leaves on a tree. They could be really tiny, like grains of sand. There could simply be too many of them. What's the highest number you've actually counted up to? I'm not sure if I've ever kept going beyond about two hundred. During one phase of insomnia I did try counting sheep (nothing else worked, so I thought: why not?) but I don't think I got beyond two hundred – I got bored of counting long before getting sleepy.

In normal life we definitely can't count anything that is infinite, because we definitely won't ever get there, and not just because we're bored. But in mathematics, counting doesn't mean actually saying the words "one, two, three, four..." out loud. It means matching up the objects you're counting with the objects in an official number bag. And here's the crux of the matter:

- The official number bag for 1 contains 1 object: the 0 bag.
- The official number bag for 2 contains 2 objects: the 0 bag and the 1 bag.

* The official number bag for 3 contains 3 objects: the 0 bag, the 1 bag, and the 2 bag.

Now once we know how to make the n bag, we can make the $n + 1$ bag: we just take the 0 bag, 1 bag, 2 bag, all the way up to the n bag, and put them in an official bag.

What we have done here is redefined the way of getting from one number to the next one, the process of adding one. We have redefined this to be "putting all previous bags into a new giant bag." A dramatic drumroll is in order at this point because we now have the key to infinity, and all that remains is to use it to unlock the secret. Here's how.

We have defined all the natural number bags, and we'll just keep going with the giant bag process: we put all *those* bags into one new giant bag. How many bags are in this super-giant bag? There is one for every natural number, so there are infinitely many bags.

We have made the official number bag for infinity: it is the bag containing all the natural number bags.

Or rather, we have made *one* official number bag for infinity – it contains an infinite number of objects where each one is labeled by a natural number. We are soon going to see that some other infinities are bigger than this. What could that possibly mean?

Let's start by "counting up to infinity" a few times. Remember, this doesn't mean actually saying all the natural numbers out loud (life isn't long enough). It means matching objects up with the objects in our official infinity bag, by means of a bijective function. We've actually already done this a few times, by means of Hilbert's Hotel. Matching objects up with the ones in the infinity bag is the same as evacuating people into a normal one-floor Hilbert Hotel. We can't evacuate two people into the same room, and we don't want to waste any rooms.

In Chapter 2 we did various evacuations. We saw that we could evacuate a two-floor hotel, a three-floor hotel, and even an infinite-floor hotel. Instead of actually performing the evacuation, we could have matched up the people with the objects in our infinity number bag, just to see if it was possible. This is the mathematical notion of countability. An infinite set is called "countable" if it is possible to match its objects up with the objects in our official infinity bag. But the objects in the official bag are just the natural numbers. So what we're saying is that an infinite set is called countable if the objects can be matched up with the natural numbers. Or, formally, if there is a *bijective function* to the natural numbers. This kind of infinite set is called *countably infinite*. Yes, we are going to see some infinite sets that are *uncountably infinite*.

Now that we are starting to get the hint that there are different types of infinity, we should start being more careful how we write infinity. The symbol ∞ just means any old non-finite thing. However, we now have a very specific notion of infinity, which corresponds to our official infinity bag containing all the natural numbers. Mathematicians sometimes call this ω, the Greek letter omega. Omega is the last Greek letter, but ω is just the beginning of a whole series of bigger and bigger infinities.

Surprisingly Countable Sets

On the one hand, you might think it's impossible for anything to be bigger than infinity. On the other hand, if we throw some more things into the giant bag of natural numbers, won't it have more things in it? This is where we have to start really rethinking our intuition about "more" and "less," because where infinity is concerned, things don't quite work like that. If we start trying to find infinite sets of things that cannot be matched up with our ω bag, we'll find it's rather hard. We can do all sorts of things that seem to make the bag bigger and yet find that it's still countable.

Let's start by taking the set of all the natural numbers and making it "bigger" by putting something else in, say an elephant.

We now want to see if this is countable – can we match these objects up precisely with the natural numbers?

We could try matching all the natural numbers in the first set with themselves. But then where will the elephant go? That was a bad plan. Instead, we should match the elephant up with the number 1, and then match all the actual numbers up with the number *one* bigger, so that 1 is paired with 2, 2 with 3, and so on. This way everything, including the elephant, is paired with a natural number, and all the natural numbers (in the official bag) get used.

This is just like when a new guest arrives at Hilbert's Hotel and all the other guests have to move up one. And what this is saying is that if you put one extra object into an infinite set, it doesn't actually get any bigger. This is one of our basic intuitions about infinity, that "infinity plus 1 is infinity." We have now made rigorous mathematical sense of it.

The two-floor Hilbert Hotel was two sets of natural numbers, and we've already shown that this has a bijective function to the natural numbers. This says that if you take two countably infinite sets and throw them all in together, the giant resulting set is still countably infinite – that is, you still don't have "more" things. We have made sense of the fact that "two times infinity is infinity."

The infinite-skyscraper Hilbert Hotel was a countably infinite number of sets of natural numbers. And yet, we could still evacuate everyone into a one-floor hotel. This shows that even if you take a countably infinite number of countably infinite sets, you still don't really have "more" things. In fact we have made sense of the fact that "infinity times infinity is infinity."

Instead of thinking about impossible hotels, we can think about different kinds of numbers. What about the integers? That's basically twice as big as the natural numbers, because there are the positive numbers *and* the negative numbers. Surely there are more integers than natural numbers? And yet, if we had a Hilbert Hotel with negative room numbers as well as positive room numbers, we could still evacuate it into a normal one-floor Hilbert Hotel because it's really no more difficult than a two-floor Hilbert Hotel, which we've already done. Instead of some rooms being on the second floor, they've been labeled with negative numbers, as if there's an east wing and a west wing.

What about the rational numbers? Surely there are more rational numbers than natural numbers? They get so close together on the number line. In fact, if we had a Hilbert Hotel with rational-numbered rooms, we could still evacuate it, although it would be a bit complicated and involve several stages.

First of all, remember that every rational number can be written as $\frac{a}{b}$ where a and b are integers and b is not 0. Now we are going to start by dealing with the positive rational numbers. We will initially evacuate all these people into an infinite-skyscraper hotel, by taking the person in room $\frac{a}{b}$ and evacuating them to room a on floor b of the skyscraper hotel. Pretty clever, eh? From there we can evacuate them to a normal hotel. What about the negative rooms? Well, we could evacuate them to another skyscraper hotel, and from there into a one-floor hotel. Then we'd be back to having two one-floor hotels, and we know perfectly well how to fit those into one hotel. This is a very typical mathematical process of turning new problems into old ones that we already know how to solve.

You might have noticed that when we go from the rational hotel to the skyscraper hotel, some rooms in the skyscraper hotel will remain empty. For example, $\frac{1}{2}$ is the same as $\frac{2}{4}$, so once we've put the guest from room $\frac{1}{2}$ into room 1 on floor 2, we have nobody going into room 2 on floor 4. This means our function isn't bijective, but in fact that doesn't matter, as long as it's injective. That is, as long as you can evacuate everyone into their own room it doesn't matter if some rooms are left empty; you could always get everyone to move backward into the empty rooms afterward.

So the integers and the rational numbers are still no bigger than the natural numbers, despite it really seeming like there are more of them. But this doesn't mean that *all* infinite sets are the same size. You might notice that we haven't yet examined how many real numbers there are. If we throw the irrational numbers into the bag along with the rational numbers, what happens? The answer is we now really do have more objects. In the next chapter we are going to use this to justify that implausible claim, that some things are more infinite than others.

If you see a churning river with swirling white water do you feel the urge to find a raft and jump in? I don't if it's a real river, but I do if it's swirling mathematical waters. I don't particularly like being physically tossed around, but I do like feeling my mind spin as I struggle to get my head around something that seems impossible at first. The fact that some things are more infinite than others is a piece of mind-spinning white-water mathematics, and we're now ready for it.

6 Some Things Are More Infinite than Others

Children sometimes have this kind of argument:

"I'm right."
"I'm more right."
"I'm right times a hundred."
"I'm right times a million."
"I'm right times a billion."
"I'm right times infinity!"
"I'm right times two infinity."
"I'm right times infinity squared!"

However, we seem to have discovered that "two infinity" is no bigger than infinity, and even infinity squared, which is infinity times infinity, is also no bigger than infinity. So in the end, the children are not becoming any more right than each other. Is there anything they can do to beat each other in this argument once one of them has declared that they're right times infinity? There is: they just need to find an infinity that's more infinite than the other child's infinity! We'll now see two ways that things can be "more infinite" than the natural numbers. The first involves trying to count the irrational numbers.

There Are More Irrational Numbers than Rational Numbers

There are more irrational people than rational people in the world, probably. In fact, most people are a bit irrational – that, to me, is an important aspect of being human and not a computer. It is extremely difficult to be completely, totally, and utterly rational, only using logic ever in your whole life. Emotions aren't rational, taste isn't rational. Surely you like some food more than others. Occasionally that can be explained rationally: for example, I don't like hot chilis because they hurt my mouth. But why don't I like cinnamon? I don't know; I just don't like it. And when people discover this they quite often react as if that is completely irrational. I admit that is my gut reaction to people who don't like chocolate before I remind myself that it's just a matter of taste, and that I know more people who dislike chocolate than people who share my dislike of cinnamon. Anyway, even if you don't think of yourself as actively irrational, you're surely not purely rational. Our everyday use of these words leaves more room for gray areas than our mathematical use does; in fact one of the whole aims of mathematics is to eliminate gray areas, not from the whole world, which would be impossible and undesirable, but from whatever we're thinking about right now.

There are also more irrational numbers than rational numbers in the world. It is in fact quite difficult for a number to be rational. Being a ratio of two integers is a remarkable coincidence. If this sounds bizarre to you, it's probably because I'm distinguishing between things that are *mathematically likely* and things that are *humanly likely*.

If someone stops you in the street and asks you to think of a number, you are almost certain to think of a rational number. In fact, I'd guess that you'll probably think of a positive whole number. (Although a certain type of person will say "π" to make a point.) This doesn't mean that there are more positive whole numbers than any kind of number; it just means that our brains are more familiar with those numbers and will tend to gravitate toward them. They are called the "natural" numbers after all.

However, if I now ask you to draw a circle with your chosen number as its radius (say in centimeters), the area of your circle is *almost certain* to be irrational. Remember, the area of a circle of radius r is πr^2. The number π is definitely irrational, so if your radius is a whole number, or any rational number, your area is doomed to be irrational.

> If you multiply a rational number by an irrational number, the result is irrational. Likewise, if you add a rational number and an irrational number, the result is irrational. However, if you multiply two irrational numbers, the result could go either way. For example, $\sqrt{2} \times \pi$ is still irrational, but $\sqrt{2} \times \sqrt{2} = 2$, which is rational.

Can you try to think of a radius that would make the area of a circle rational? How are we going to get rid of that pesky π that's floating around in the formula? We're going to have to cancel it out somehow. If we pick the radius r to be $\frac{1}{\sqrt{\pi}}$, then πr^2 will neatly cancel out to be

$$\pi \times \left(\frac{1}{\sqrt{\pi}}\right)^2 = 1.$$

I'm guessing that if I stopped you in the street and asked you to think of a number, then $\frac{1}{\sqrt{\pi}}$ would not be your answer. (Although now that I've suggested it, perhaps it will be in the future.)

What about other rational areas? Remember that a rational number is one that can be expressed as $\frac{a}{b}$ where a and b are both integers and b isn't 0. So let's try and make this fraction the area of a circle. If we put

$$r = \sqrt{\frac{a}{\pi b}}$$

then we get

$$\begin{aligned}\pi r^2 &= \pi\left(\sqrt{\frac{a}{\pi b}}\right)^2 \\ &= \frac{\pi a}{\pi b} \\ &= \frac{a}{b}.\end{aligned}$$

The upshot is there are tons more irrational numbers than rational numbers out there, it's just that the rational ones are much more prominent in our thoughts most of the time. However, there was nothing very rigorous about the argument. I vaguely threw around words like "probably" and "almost certain" and "tons more." Often in mathematics we start by thinking about how things feel, using fuzzy language like this, before starting to hone it into a logical mathematical argument. The honing is what we're going to do next.

So Many Irrational Numbers

We're going to "count" the irrational numbers, except that we don't have a way of saying what irrational numbers *are*. Irrational numbers are defined by what they're not: they're the real numbers that are not rational. We do have a way of saying what real numbers are (although we will have to be a bit vague about it until Chapter 15), and what rational numbers are. So it's easier to "count" the irrational numbers by a process of exclusion, rather than by counting them directly. We're basically going to say, "Wow, there are a lot of real numbers. And not many are rational. So if I throw away the rational ones, there's still tons left that are irrational." We're just going to be a bit more precise (or technical) than that.

Instead of showing that the irrational numbers are uncountable, we're going to show that the real numbers are uncountable. We already know the following things:

1. The rational numbers are countable.
2. If you put two countable sets together, you get another countable set (as in the two-floor Hilbert Hotel).

This tells us that *if* the irrational numbers were countable, the real numbers would have to be as well. But we'll show that real numbers are *not* countable, so the irrational numbers can't be countable either.

This is a bit like dominant and recessive genes. Imagine that being uncountable is like a dominant gene, and being countable is like a recessive gene. So if you put a countable and an uncountable set together, you get an uncountable one, because that's the dominant gene. Now if you put a countable set together with an unknown set (the irrationals), you can deduce what the unknown one was if you know the outcome.

rationals	**if reals**	**then irrationals**
countable (known)	countable	countable
	uncountable	uncountable

The Real Numbers Are Uncountable

The real numbers are difficult to pin down, but for now let's say they are all the possible decimal numbers, where the decimals are allowed to go on forever, repeating or not repeating. (In fact all decimals go on forever if you put 0's on the end; we just don't usually bother writing all those 0's.) We are so used to thinking about decimal expansions that go on forever, it might be difficult to see why this is not a perfectly good definition of the real numbers. We can't really talk about this until we've thought about infinitesimally small things, in Chapter 14.

We're going to show that these never-ending decimals are uncountable using a clever trick that Georg Cantor came up with, which is now known as the *Cantor diagonal argument*. In fact, we're going to show that the real numbers between 0 and 1 are uncountable all by themselves. (Restricting our attention to these numbers makes the argument more streamlined.)

Let's put this question another way. We're going to think about evacuating huge hotels again. In Chapter 2 we mentioned the idea of the real-number hotel, where there is a room for every real number. We're now going to do something more modest: a hotel that "just" has a room for every real number from 0 to 1. I say "just," but this is still a super-giant hotel, and it's the first one we will show cannot be evacuated into the ordinary Hilbert Hotel.

It's convenient only to think about the numbers between 0 and 1 because it saves us worrying about the part of the number before the decimal point: every room has a different decimal expansion that starts 0.something and goes on forever. (Let's not think about how long it would take you to ask for your key.) Now imagine trying to evacuate this hotel: this is the question of finding a perfect pairing between the real numbers (from 0 to 1) and the natural numbers, which is what we mean by "countable."

We could start by trying a few things.

Can we just move people into their new rooms in order of size? So the smallest decimal-number room moves into room 1, and so on? This won't work, because there is no smallest decimal number, just like there's no biggest number. Is 0.0000000000001 the smallest possible decimal number? No, because we could always insert a few more 0's in there to make it smaller.

Can we start with the decimals that only have one decimal place, and then move on to the ones with only two decimal places, and keep going like that? After all, there's only a finite number of each of these: there are only 10 rooms with one decimal place, namely $0.0, 0.1, 0.2, 0.3, \ldots, 0.9$. How many are there with two decimal places? Well we have 10 choices for the first place and 10 choices for the second place, which makes $10 \times 10 = 100$ rooms. If the second decimal place is 0, it will just be the same as one of the above numbers, so we have these 100 extra rooms to deal with at this stage:

0.01	0.11	0.21	0.31	0.41	0.51	0.61	0.71	0.81	0.91
0.02	0.12	0.22	0.32	0.42	0.52	0.62	0.72	0.82	0.92
0.03	0.13	0.23	0.33	0.43	0.53	0.63	0.73	0.83	0.93
0.04	0.14	0.24	0.34	0.44	0.54	0.64	0.74	0.84	0.94
0.05	0.15	0.25	0.35	0.45	0.55	0.65	0.75	0.85	0.95
0.06	0.16	0.26	0.36	0.46	0.56	0.66	0.76	0.86	0.96
0.07	0.17	0.27	0.37	0.47	0.57	0.67	0.77	0.87	0.97
0.08	0.18	0.28	0.38	0.48	0.58	0.68	0.78	0.88	0.98
0.09	0.19	0.29	0.39	0.49	0.59	0.69	0.79	0.89	0.99

When we do the rooms with three decimal places there will be more, but it will still be finite. This seems like a promising approach, but here's the problem: we will only ever evacuate the people with decimal expansions that stop somewhere. We will never get round to the ones that go on forever. This is even if we keep evacuating "forever." It's a bit like the fact that if you keep adding one to a number, it will get bigger and bigger forever but it will never actually *be* infinite, even if you keep going forever. Likewise, if we keep evacuating in this way forever, we will keep dealing with longer and longer decimal expansions – we will keep adding one to the length of the expansion, but the expansion will never actually *be* infinite.

If you're not convinced by this argument, that's excellent, because it's not a mathematical argument, so you probably really shouldn't be convinced by it. Cantor's diagonal argument, however, is a watertight mathematical argument. We have to show that it is *impossible* to evacuate everyone to the one-floor hotel, no matter how hard we try. It sounds like this would be very difficult to prove, because it sounds like we have to try every single method of evacuation and show that it fails. The thing that's clever about Cantor's argument is that you don't actually go through every process of evacuation. You just assume that it has been done, that everyone is now in the one-floor hotel, and you derive a contradiction. We are going to show that if some smarty-pants *claims* to have evacuated every room, we will always be able to find at least one person who was left out. Mathematically this means that if we claim to have matched up all the decimals with natural numbers, there will always be at least one decimal number that did not get assigned to a natural number.

Here's how it works. You go to room 1 and knock on that person's door. You ask them what their old decimal room number was. Actually, you don't ask them for the whole expansion, you just ask them for the first digit, like when you're logging on to online banking and they just ask you for the third and seventh letters of your password. Anyway, after they tell you their digit, you add 1 to it and write it down. So if their old first digit was 3, you write down 4. If it was 8, you write down 9. If it was 9, you don't write down 10 (because that's not a digit), but you write down 0. This sounds mysterious, but bear with me.

Now move on to room 2, and ask that person for the 2nd digit of their old room number. Again, add 1 and write it down next to the previous thing you wrote down.

Now move on to room 3, and ask that person for the 3rd digit of their old room number. Again, add 1 and write it down next to the previous one.

What we're doing here is building up a new decimal-expansion room number. You will get the *n*th digit by asking the person in the new room *n* for the *n*th digit of their old room number, and adding 1.

Now if you did this in "reality," you would never get to the end, but the mathematical argument doesn't involve physically knocking on doors. The number that we're building up here exists regardless of whether or not we have time to write it down, just like the planet Neptune existed before humans knew it was there.

The question is: Where is the person who used to be in this room, the one with the decimal expansion we just derived by this business of adding 1 to digits? Which natural-number room have they been evacuated into?

They can't be in room 1, because the new decimal number differs from that person's old room number in the 1st digit. They can't be in room 2, because the new decimal number differs from that person's old room number in the 2nd digit. Nor can they be in room 3, 4, 5, or any room number *n*, because the decimal number differs from that person's old room number in the *n*th digit. So we have found a person who did not get evacuated anywhere. This is the contradiction, and shows that the smarty-pants was wrong: they have not successfully evacuated everyone. This argument works for *any* evacuation they claim to have done, which shows that the evacuation is not possible.

The reason this is called a diagonal argument is because if we write down the decimals in a grid, we will be looking down the diagonal. Let's suppose that the first few evacuated rooms look like this:

new room no.	old room no.
1	0.238795317···
2	0.984718573···
3	0.389716438···
4	0.777362889···
5	0.444317895···
6	0.879000001···
7	0.892225673···
8	0.191919234···

Then the digits we ask for when we knock on each person's door are the ones in bold here, along the diagonal.

new room no.	old room no.
1	0.**2**38795317···
2	0.9**8**4718573···
3	0.38**9**716438···
4	0.777**3**62889···
5	0.4443**1**7895···
6	0.87900**0**001···
7	0.892225**6**73···
8	0.1919192**3**4···

In this case the new number we'll produce will start like this

0.39042174 · · ·

and we can show that this person is not in any room *n* by looking at their *n*th digit, which is different from the person who is actually in room *n*. So this person has not been evacuated to any room, and the smarty-pants has failed.

This shows that the real numbers between 0 and 1 are uncountable: if you try and make a perfect pairing with the natural numbers, at least one of the real numbers is doomed to be left out. We can think of this as meaning that the real numbers are "more infinite" than the natural numbers.

This way of proving something by defeating a smarty-pants who thinks they can do something is sneaky and useful. The smarty-pants will come back to try us later, when we're thinking about infinitesimally small things.

Decision Fatigue

Another way in which something can be "more infinite" than the natural numbers is more subtle, and perhaps relates more to the word "uncountable." We can't evacuate the real-number hotel into the natural-number hotel because someone is doomed to be left out in the cold. However, another way in which we can fail to evacuate a hotel is if we get decision fatigue. Imagine you have a hotel full of double rooms and you need to evacuate it into a hotel of single rooms. You could evacuate the two people in room 1 to rooms 1 and 2, and the two people in room 2 to rooms 3 and 4, and so on, with the two people in room n to rooms $2n$ and $2n - 1$. On the face of it, this sounds much like evacuating a two-floor hotel into a one-floor hotel, but how would you write out the instructions this time?

> "If you're currently in room n, then one of you go to room $2n$ and the other to room $2n - 1$."

What if they say, "Which of us should go in which room?" You could say, "Whoever's oldest should go in room $2n$, and the younger person should go in room $2n - 1$." But what if all the even-numbered rooms are better and the younger person of each pair gets upset? Or – bear with me – what if all the pairs of people are exactly the same age as each other? You might ask them just to decide for themselves, as it doesn't really matter that much. But what if they're very indecisive and need you to decide for them? Evidently this analogy is breaking down somewhat, but the idea of being able to write down instructions is supposed to be that the people simply follow the instructions without having to use any thought. If they have to make a decision along the way, then mathematically this is not an unambiguous set of instructions. To put it another way, it is not a process that a computer could follow, also known as an algorithm.

This situation is often explained as the difference between counting shoes and counting socks. Suppose you had an infinite number of pairs of shoes. That is, a countably infinite number. So you can line them all up in shoe boxes labeled 1, 2, 3, and so on. Does this mean that the individual shoes are countable? What if you had an infinite number of pairs of socks?

I try not to count my shoes too often, as I might be shocked by how many pairs I have. My excuse is that my feet are really big and it's hard to find shoes that fit and that aren't ugly. When I was fat, my feet were even bigger and I got into the habit of buying every pair of shoes that fit, wasn't ugly, and wasn't too expensive. In those days not many shoes met those criteria, so buying them all was quite feasible. I basically had an algorithm for buying shoes, and it still meant that I only owned about four pairs of shoes when I graduated. (This is still more than some men I know.) When I lost weight, I discovered that my feet got smaller as well (I did not realize that feet could get fat). The difference was only about half a size, but this critical half size moved my feet from out of the standard range of women's shoes into the petite range. But I still had the same shoe-buying mentality, which meant I ended up buying quite a lot of shoes. I suppose as vices go, that's better than impulse-buying sports cars. Also maybe I'm protesting too hard. I now have to make decisions when I buy shoes, rather than just following an algorithm.

However, no matter how many shoes I buy, I will still only have a finite number of pairs of shoes. But let's imagine having an infinite number of pairs of shoes, all neatly lined up in shoe boxes numbered according to the natural numbers. (My shoes are never neatly lined up.)

This means that we have a countable number of *pairs* of shoes. Do we have a countable number of shoes? That is, can we now line the individual shoes up in order, matched up with one natural number per shoe instead of one natural number per pair? Yes we could, just like the two-floor Hilbert Hotel. We could decide to start with the left foot each time, so we could go:

> left shoe 1, right shoe 1, left shoe 2, right shoe 2, left shoe 3, right shoe 3,

and so on. Written mathematically:

* The left shoe n will go to position $2n - 1$.
* The right shoe n will go to position $2n$.

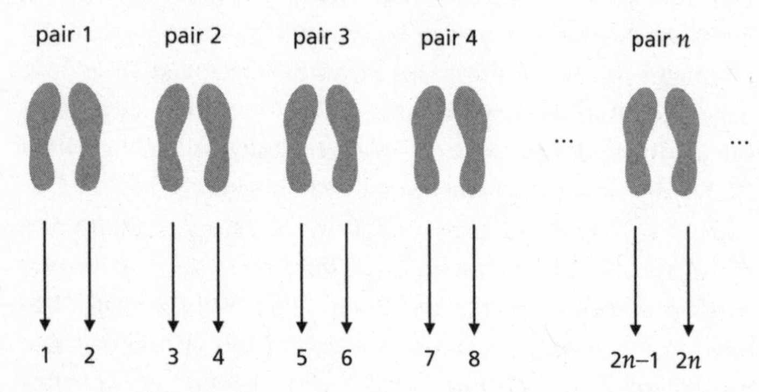

This is just like the example of the red and blue natural numbers being matched with the purple ones, where this time instead of blue we have "left shoe" and instead of red we have "right shoe."

Now let's try this for socks. Suppose we have an infinite number of pairs of socks and they're lined up like the shoes, with pair number 1, pair number 2, and so on. (I'm much less interested in socks than shoes, so all my socks are identical black ones so that I don't have to think about matching them in pairs.) Can we now line up the individual socks? At first sight this seems just the same as the shoes, but there's a problem: there's no such thing as a right sock and a left sock. Left and right socks look exactly the same. So how can we specify which one is going to go first? We can't, but we might think we can just arbitrarily pick one to go first. Of course, we'll have to make that decision for each pair of socks separately. For the shoes, it's true that I arbitrarily decided the left shoe would go first, but I only had to make *one* arbitrary decision. With the socks, every time I get to a new pair I have to make a new arbitrary decision about which sock should go first. And so we run into the problem of decision fatigue. Or at least, the mathematical version of decision fatigue.

The theory of decision fatigue is that making decisions is exhausting, whether they're small decisions ("What shall I have for breakfast?") or large decisions ("Which house shall I buy?"), and as you make more decisions in a day you get increasingly exhausted from it. Making decisions is hard. You can weigh up the pros and cons but in the end you have to make a leap – you can't just follow logic all the way to a conclusion, because then it wouldn't be a decision, it would be a deduction.

There is a mathematical version of this, just as in the example of the pairs of socks. The idea is that we can make one arbitrary choice, and we can make two or three or any finite number of arbitrary choices, but can we make an *infinite* number of arbitrary choices? In the sock example this comes down to whether or not we can write down a *formula* for how to line up all the socks in a row. Technically for the socks to be countable we have to make a function from the set of all socks to the set of natural numbers (as if we were evacuating individual socks into a Hilbert Hotel). But how are we going to say what this function is? How are we going to say which sock of each pair is going to go first, given that there are no distinguishing features?

You might think that you can just decide that you are going to pick up each pair of socks with one sock in each hand, and whichever one you have in your left hand will go first. It's a bit subtle, but in mathematics this still doesn't count as a good enough instruction. "Decision" isn't really the right word in the end: it's really about choice. Even if you pick up the socks randomly, at some level you had to choose which sock went into which hand, and you'll have to make a separate choice infinitely many times.

The question of whether or not this is possible is not exactly resolved in mathematics, but is a tricky and subtle point that still worries mathematicians. It is called the *Axiom of Choice*. An axiom is a basic assumption that you've decided to take as true without justification, a building block from which everything else is going to proceed. One way of thinking about axioms is that you're not saying they *are* true, you're just creating a world in which they are true, and studying it to see what will happen.

The Axiom of Choice says that it is possible to make an infinite number of arbitrary choices. In a world where the Axiom of Choice is true, the socks are countable. However in a world where the Axiom of Choice isn't true, the socks are *uncountable*. Likewise, when we had an infinite number of double rooms in the hotel, we could use the fact that there's an older person and a younger person (if they're not all the same age) so that we didn't have to keep choosing which of each pair of people we'd evacuate into which room.

Mathematicians don't exactly care whether or not the Axiom of Choice holds over all, but they do care whether you have to use it in any given situation or not. It is sufficiently troubling that we feel the need to point out every time we use it, a bit like trucks that start beeping when they go into reverse.

So this gives us another way in which things can fail to be countable – not because there are too many of them, but because they're too indistinguishable. This even happens to me when I'm counting people in a room, say in a classroom with a new group of students I don't know yet. I hope you've had this experience as well, so it's not just me and my apparent inability to count. If everyone is sitting in neat rows already, it's fine, because you just work your way along the rows. If everyone is sitting in randomly placed chairs, it's much harder, because you keep having to decide who to count next. I usually find it works for nine or ten people, and then I get confused about who I've already counted and who I should choose next. Imagine trying to do that for an infinite number of people. That would be extremely difficult even without the issue of running out of time in our short finite lives.

There are many jokes about mathematicians being unable to count, but the truth is that counting is a much deeper topic than we think about when we say a child has "learned to count." From me getting confused counting coffee scoops while talking, to the issue of uncountably infinite sets, thinking about these issues has been a fertile ground for mathematicians developing ideas. We have seen that there are more real numbers than natural numbers, and in the next chapter we're going to try and count them so that we can see how many more there actually are. The insight gained from doing this will take us, in the following chapter, to the hierarchy of increasingly infinite infinities and finally an answer to the children who are trying to be more right than "right times infinity."

7 Counting Beyond Infinity

I recently visited Tent Rocks National Monument near Santa Fe, New Mexico. I went on a dramatic hike up through the conical rock formations in a narrow canyon, up to a ridge at the top of the mesa with spectacular views of the Rio Grande Valley. It was satisfying physical exertion, had breathtaking close-up views of tent rocks all the way up, and culminated in the dramatic arrival at the ridge with the sweeping views off to the horizon. I had a few moments of fear on the way up as I'm not terribly brave and the view of the drop down the rock face was a bit much for me on occasion, especially when the part we were walking on was particularly steep with loose rocks making me feel that I didn't have a good foothold on anything. When I got to the top I enjoyed a brief sense of achievement and then realized there was more. It wasn't a peak but a ridge, so we could keep walking along it, now with the drop down the rock face on *both* sides. I thought about continuing but decided that was enough bravery for me, and I went down again.

You might have felt like this on our climb to infinity. Perhaps you felt that you didn't have a strong foothold on anything anymore, and now that we've finally made it to infinity you'd like to leave it at that. I'm much braver with mathematics than I am with rock faces, so we're now going to keep going. We've shown that the real numbers are more infinite than the natural numbers, but how much more infinite are they? How far is it from the one infinity to the bigger one? We're going to count.

Of course, we can't count the real numbers by listing them and seeing how many there are. Instead, we're going to count them as we counted things in the previous chapter: abstractly. We're going to think about how they're constructed and come up with a relationship between them and the natural numbers, to see how far apart they are. First we're going to build up to this by counting some smaller sets to get used to how this abstract method of counting is done.

Abstract Counting

Imagine a prix fixe menu with choices like this:

Soup of the day

∞

Grilled salmon with lemon beurre blanc

or

Herb-roasted chicken with mashed potatoes

∞

Chocolate cake

or

Lemon tart

How many possible dinners are there? We could count them in the concrete way, by writing out a list:

1. salmon and cake
2. salmon and tart
3. chicken and cake
4. chicken and tart

This is not too arduous for a menu as short as this, but imagine if it were a ten-course meal with three choices for every course – it would take a rather long time to count all the possible combinations in that case. This is where abstract counting starts to come in handy. We count by reasoning rather than by lining things up against the numbers 1, 2, 3, 4, and so on. As an intermediate step, we can draw a little tree of possibilities like this:

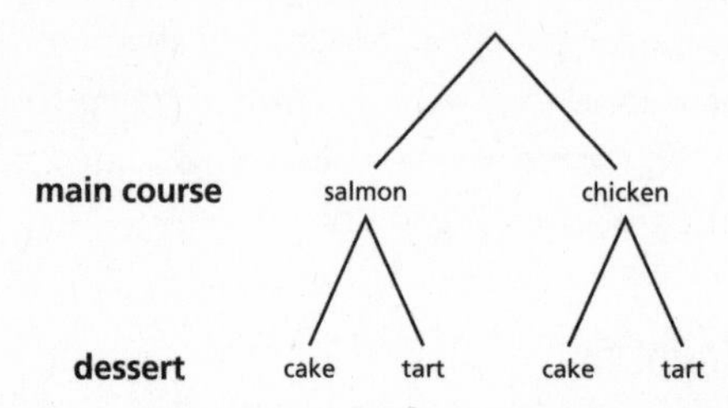

The top level of this (upside-down) tree says we have two possible choices for the main course. The bottom level says that no matter which of those we choose, we proceed to have two possible choices for dessert. Each path down through the tree gives us one possible menu combination, and there are four paths. But we can also see this abstractly by seeing that the two choices on the top level get multiplied by the two choices on the bottom level, so the total number of choices is $2 \times 2 = 4$. At this point it's possible to get confused about whether we should add or multiply our 2's, because $2 + 2$ would also give us the same answer. But let's try it with a choice of starter as well.

Soup of the day
or
Green salad
∞
Grilled salmon with lemon beurre blanc
or
Herb-roasted chicken with mashed potatoes
∞
Chocolate cake
or
Lemon tart

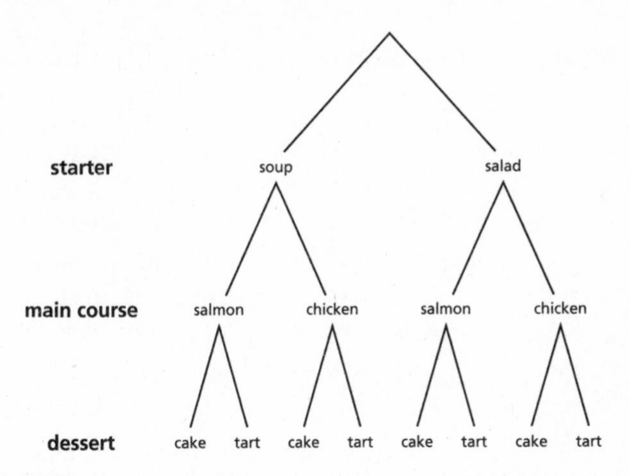

This time the two choices on the top level get multiplied by the two choices on the second level and then this gets further multiplied by the two choices on the bottom level. So the total number of possible combinations is $2 \times 2 \times 2 = 8$.

We can do something similar to count decimal numbers. We're going to build up to counting all the real numbers by first just thinking about the numbers between 0 and 1, like we did in the previous chapter. We're basically going to use the method that didn't quite work that time, when we tried to evacuate the decimal-number hotel decimal place by decimal place. We start by looking at the first decimal place, which is like the first course on the menu. Except now instead of there being just two choices, there are ten choices, as there are ten possible digits that could be in the first decimal place. The ten possible numbers with only one decimal place are these:

$$0.0, 0.1, 0.2, 0.3, 0.4, 0.5, 0.6, 0.7, 0.8, 0.9$$

For numbers with two decimal places, it's like a two-course menu, where there are ten choices for each course. So there are $10 \times 10 = 100$ possible combinations. In terms of branching trees, the first level here would have 10 possible branches, and then each of those branches would have 10 branches attached, making 100 leaves in the end, or 10^2. We saw in the previous chapter that the 100 possible numbers with two decimal places are:

0.00	0.01	0.02	0.03	0.04	0.05	0.06	0.07	0.08	0.09
0.10	0.11	0.12	0.13	0.14	0.15	0.16	0.17	0.18	0.19
0.20	0.21	0.22	0.23	0.24	0.25	0.26	0.27	0.28	0.29
0.30	0.31	0.32	0.33	0.34	0.35	0.36	0.37	0.38	0.39
0.40	0.41	0.42	0.43	0.44	0.45	0.46	0.47	0.48	0.49
0.50	0.51	0.52	0.53	0.54	0.55	0.56	0.57	0.58	0.59
0.60	0.61	0.62	0.63	0.64	0.65	0.66	0.67	0.68	0.69
0.70	0.71	0.72	0.73	0.74	0.75	0.76	0.77	0.78	0.79
0.80	0.81	0.82	0.83	0.84	0.85	0.86	0.87	0.88	0.89
0.90	0.91	0.92	0.93	0.94	0.95	0.96	0.97	0.98	0.99

We can now generalize this to any n decimal places, which would be a menu with n courses or a tree with n levels. For numbers with n decimal places, there are $10 \times 10 \times \cdots \times 10$ possibilities, where that 10 appears n times. That's 10^n.

Now for a slight leap of faith: for numbers with an infinite number of decimal places, there are "ten to the power of infinity" possibilities. I say that this is a slight leap of faith, but really it's some more set theory. We are trying to count the real numbers, which really means we're trying to find an official bag of objects that matches up with the real numbers. You might think we could just say "it's the set of real numbers" and be done with it, just like we did with the natural numbers. But it would be more helpful and illuminating to be able to relate it to the natural numbers. "Ten to the power of infinity" is not far off, but to interpret this as a set of things, and to get a more satisfying answer, it's better to do this in binary.

Digression on Binary

Binary is a system of representing numbers where instead of using the digits 0, 1, 2, 3, all the way up to 9 as we do in the decimal system, we only use the digits 0 and 1. It's amazing how much information you can encapsulate using only the digits 0 and 1. Computers, basically, run entirely on this system, as if everything is just an on/off switch, and there are millions and billions of them. You can get very many different configurations from a very small number of on/off switches.

With two switches, you can get four configurations:

switch 1	switch 2
off	off
off	on
on	off
on	on

With three switches, you get eight configurations:

switch 1	switch 2	switch 3
off	off	off
off	off	on
off	on	off
off	on	on
on	off	off
on	off	on
on	on	off
on	on	on

This is just like our two- or three-course menu options. With the switches the tree now comes out like this:

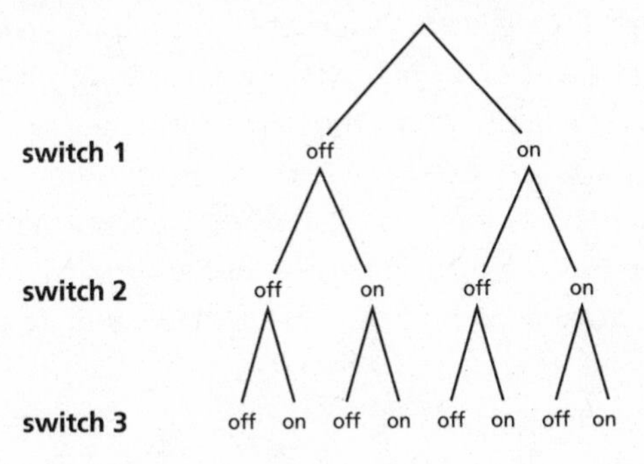

Each of these eight ending points gives us one of the eight configurations for three switches; we just have to follow the path down from the top of the picture to the given ending point and read off whether we passed through "off" branches or "on" branches on the way down.

We've been calling these pictures trees because that's what they're called in mathematics. It's slightly silly because trees in normal life usually start at the bottom and grow upward, but trees in mathematics often start at the top and grow downward, mostly because this is how we read down the page. Sometimes people draw them sideways:

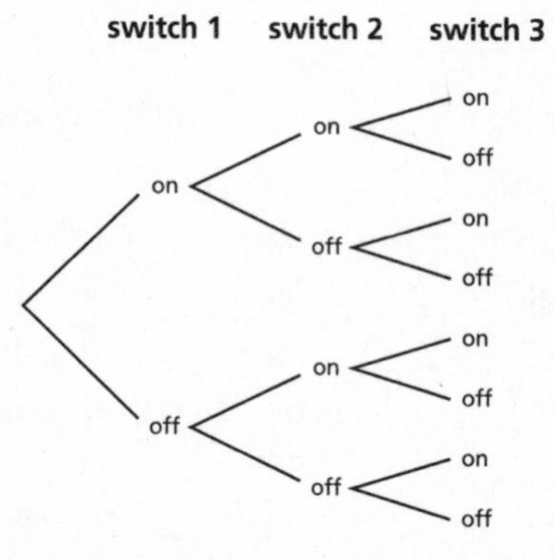

I hope you can see that it doesn't really matter which way up we draw a mathematical tree, because whichever way up the diagram is, the information we're encoding is still the same. We might call this a *schematic* diagram, as it's depicting a scheme of how things fit together abstractly, rather than how they fit together physically. As mathematics gets more abstract, diagrams become more and more prominent as the ways that things fit together abstractly become both more subtle and more important. Moreover, the diagram often sums up the situation more succinctly than the explanation in words, just like our diagram of how to evacuate the infinite-skyscraper Hilbert Hotel.

Most of basic mathematics happens in straight lines. For example, a sum like this:

$$3 + 2 = 5$$

or an equation

$$2x + 3 = 7.$$

The symbols all sit in a row quite happily. Even if we solve that equation, we write a series of rows:

$$2x = 7 - 3$$
$$2x = 4$$
$$x = 2$$

Later on we'll see that as mathematics gets more advanced it gains dimensions. If the objects we're thinking about have shape, then there are more ways we can fit them together than just juxtaposing them in a straight line, a bit like fitting jigsaw pieces together or building something complex out of Lego. Imagine trying to explain to someone in words how to build a car out of Lego, rather than having instructions in diagrams. Just as a picture supposedly paints a thousand words, a mathematical diagram can explain a situation a lot more quickly and vividly.

From the tree diagrams perhaps you can now see that every time we add a new switch, we'll have to split every single ending point into two branches at the next level. Incidentally, the ending points are often called "leaves," seeing as they're at the ends of "branches," even though the tree is upside down.

Every time we add a switch we double the number of possible configurations. This means that for *four* switches we have $2 \times 2 \times 2 \times 2$ possible configurations, that is, 2^4, and for n switches we have 2^n possible configurations. This is just like when we were counting the decimal numbers with n decimal places, except now we have only two choices for each level instead of ten, just like for the menus.

The binary number system works like a series of on/off switches. It's just like the decimal number system except that instead of having units, tens, hundreds, thousands, etc., we have units, twos, fours, eights, and so on. Thinking about whole numbers in binary is more well known than fractions. We can compare, say, four-digit numbers in decimal and in binary like this.

	1st digit	*2nd digit*	*3rd digit*	*4th digit*
decimal	$\times 10^3$	$\times 10^2$	$\times 10$	$\times 1$
binary	$\times 2^3$	$\times 2^2$	$\times 2$	$\times 1$

So in decimal, the number 1101 can be expanded like this:

$$(1 \times 1000) + (1 \times 100) + (0 \times 10) + 1$$

whereas the binary number 1101 is expanded into decimal like this:

$$(1 \times 8) + (1 \times 4) + (0 \times 2) + 1$$

or 13, in decimal.

In decimal, a four-digit number can express anything up to 9999, which is $10^4 - 1$. Whereas in binary, a four-digit number can only express things up to $2^4 - 1 = 15$.

This might make it seem that binary is a bit feeble, especially as I promised in Chapter 5 that binary would enable us to count up to 1023 on our fingers. It's a trade-off. With binary, you can encode everything using simple on/off switches, whereas with decimal, you need each "place" to be able to give ten possible different pieces of information, that is, the digits 0, 1, up to 9. In some situations (like computers), you have a big capacity for the number of digits you can use but a small capacity for the number of different positions they can be in. Whereas, for example, in ISBN codes on books, you have limited space to print the code but plenty of different things you can print in each position. HTML color codes are even more space-saving, and are written in hexadecimal, that is, base 16, which means every position has 16 possible digits. The digits used are 0, 1, 2, 3, 4, 5, 6, 7, 8, 9, A, B, C, D, E, F.

One of my favorite uses of binary is for birthday candles. If you have seven candles, it might look like you can only deal with birthday cakes for children up to the age of seven. But if you use them in binary, with a lit candle representing "on," or 1, and an unlit candle "off," or 0, you can celebrate everyone up to the age of 128. Which is everyone on earth, currently.

And it is true that we can use binary to count on our fingers up to 1023, that is, up to $2^{10} - 1$. Here's how. Each finger has two possible positions: up or down. Up is for 1 and down is for 0. Now we have ten digits (actual digits!), and we can express all the numbers from 0 to 1023 using ten digits in binary. Here are pictures of the numbers 0 to 31, which can all be done on one hand.

Here are the corresponding numbers in five-digit binary.

00000	00001	00010	00011	00100	00101	00110	00111
01000	01001	01010	01011	01100	01101	01110	01111
10000	10001	10010	10011	10100	10101	10110	10111
11000	11001	11010	11011	11100	11101	11110	11111

It's pretty satisfying but takes some concentration, unlike the normal way of counting to 10 on your fingers. So unfortunately this method is unlikely to help free up your mental space. I'm quite sure I couldn't get very far with this while holding a conversation with someone. (I just tried, and only got up to 10 before making a mistake.)

Actually, if you really concentrate and have good control over your fingers, you could use your fingers in base 3, which means each position needs three possibilities. This requires you to be able to bend each finger halfway, independently of all other fingers, so that each finger can take three possible positions and thus represent 0, 1, or 2 in that position. Try holding up your ring finger by itself and then getting your middle finger to distinguish between half-bent and fully down. It's a bit tricky.

That was all about whole numbers, but we can use binary for fractions as well as for whole numbers. We just have to remember what decimal places really mean, and translate them into "binary places." We also have to remember not to use the word "decimal" all over the place when we mean "fraction." (I am always tempted to say "binary decimal," but that doesn't make sense linguistically.)

When we do ordinary decimal fractions, the places after the decimal point are tenths, then hundredths, then thousandths, and so on. For example, this number

$$0.3526$$

is really

$$\left(3 \times \frac{1}{10}\right) + \left(5 \times \frac{1}{100}\right) + \left(2 \times \frac{1}{1000}\right) + \left(6 \times \frac{1}{10000}\right).$$

In binary fractions, the first place is halves, and the next is quarters, then eighths, then sixteenths. So the binary number

$$0.1101$$

translates as

$$\left(1 \times \frac{1}{2}\right) + \left(1 \times \frac{1}{4}\right) + \left(0 \times \frac{1}{8}\right) + \left(1 \times \frac{1}{16}\right)$$

which in normal decimal fractions is 0.8125. We could depict the possible binary fractions in a tree, just like we did for the on/off switch configurations.

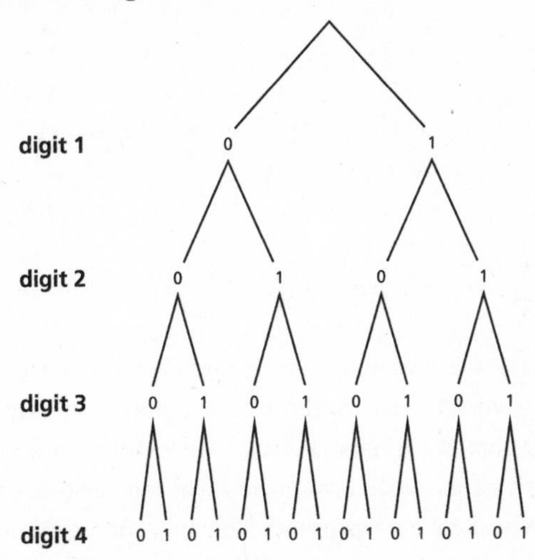

In this picture, instead of "off" and "on" branches we now have 0 and 1 branches, a bit like how we can use unlit and lit candles to represent 0 and 1 on a binary birthday cake. At the moment we're only doing four digits, so we have four levels of branching. We could have done this for ordinary decimals, but in that case we would have needed ten branches coming out of every previous ending point at every level, quickly making the tree unmanageable to draw.

Now each leaf represents a binary fraction, and we can see which one it is by following the path down from the top of the tree to that leaf, reading off the 0's and 1's on the branches as we go. So the first leaf is 0.0000, the second is 0.0001, and so on. Here they all are:

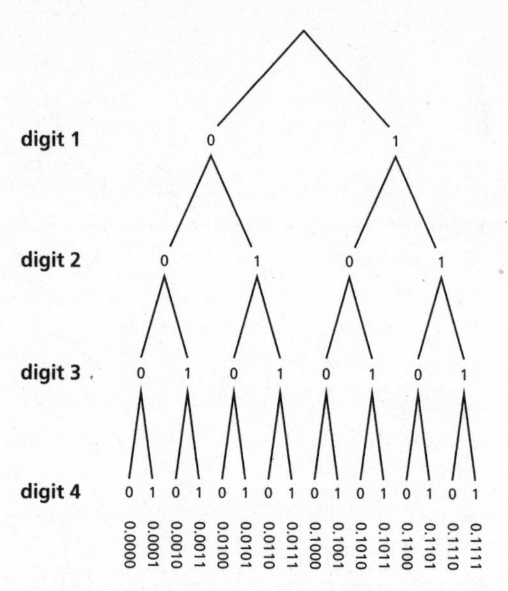

For five digits we'll have a tree with five levels, for *n* digits we'll have a tree with *n* levels, and for binary fractions with infinitely long expansions we'll have a "tree with infinitely many levels." This is a bit of an odd concept because the resulting numbers were supposed to be represented by the ending points of the tree, the leaves. But if the tree goes on forever, then there won't be any ending points. For this reason it makes more sense to think of the numbers as paths through the tree. We previously saw that we could pick any leaf and see what number it represents by following the path to it from the top of the tree. Even if there aren't exactly any ending points (because the tree goes on forever), we can think about paths through the tree, where the paths also go on forever.

Just as ordinary binary numbers can represent all the numbers that the normal decimal system can do (just using more digits than usual), binary fractions can represent all possible decimal fractions, also using more digits. So the paths through the infinite tree correspond to all the "decimal numbers that go on forever," that is, all the real numbers. Or at least, the real numbers from 0 to 1. We now can now make the following leap of faith:

* A tree with n levels has 2^n paths through it.
* A tree with infinite levels has "2 to the power of infinity" paths through it.

We still haven't dealt with the potential whole-number part of the fractions, but in the next chapter we'll see why that part doesn't matter very much. We'll also see why it's more satisfying to do this in binary rather than decimal numbers, and why it was acceptable to ignore the whole-number part.

You might think we haven't achieved much here, because we're stuck with the question: How big is 2 to the power of infinity? We are going to build up to answering this question in the next chapter. But first we're going to take in a breathtaking piece of scenery we're now in a position to admire. We have seen how to construct real numbers from natural numbers using trees. We have seen that this makes a bigger infinity. All we have to do is repeat this process and we have made a hierarchy of infinities. This is the subject of the next chapter.

8 Comparing Infinities

When children learn to climb up one step, they get very excited and climb up another and another. They marvel at how high they can go just by iterating one new thing that they have learned. If they manage to climb up a particularly big step, they will be particularly excited.

We have just learned to climb up a particularly big step: we have learned how to climb up from one infinity to a bigger one. And just like a small baby marveling at their new step-climbing ability, we too are going to climb up another step, and another. These are not physical steps, and they are not even number steps. They are steps up the ladder of infinities.

We are going to combine the following facts:

* The real numbers are more infinite than the natural numbers.
* If the infinity of natural numbers is written as ω, then there are 2^{ω} real numbers.

We are going to iterate this to show that we can keep getting more and more infinite infinities in an endless hierarchy. Of course, first we have to be clear what on earth that means.

The general question is: How should we compare the size of infinities? So far we have decided that the real numbers are *uncountable*, that is, that there is no possible way to match them up precisely with the natural numbers because some real numbers are doomed to get omitted. Intuitively this means that there must be "more" real numbers than natural numbers, but what could this possibly mean if they're both infinite? Some infinities are bigger than others – how is that possible, seeing as infinity is already infinitely big? Isn't it the biggest thing that there is? How can anything be bigger than it?

Just like questions of the soul, everlasting life, and whether or not I'm fat, this comes down to definitions. What is the definition of "fat"? In the case of infinity the question is: What is the definition of "big"? We could just give up and say it doesn't make sense, because infinity is infinitely big. But mathematics tends not to give up unless it comes up against an actual logical contradiction. If there is a phenomenon that seems to make sense *intuitively*, mathematicians are determined to find a framework in which it makes sense *logically* as well. We are stubborn like that. If I'm on a hike and the path leads to a cliff edge, I will nervously back away. But if I see a mathematical cliff edge, a sudden gap between intuition and logic, I'll eagerly go and look over the edge to see what I can do about it.

This usually involves making some careful definitions. Does this sound circular? If I want to prove that I'm not fat, I can choose my definition to be about body mass index, in which case according to most "official" definitions, I'm not fat. But if I want to prove that I *am* fat, I can choose my definition to be about waist-to-hip ratio instead, in which case by most definitions (involving risk of diabetes), I am fat.

This way of making up definitions to suit your aims might well sound a bit backward from the way that mathematics is often thought of, where you start with some objects, say numbers, and see what is true about them. For example, what happens when you add 3 and 4? What happens when you add 4 and 3? Aha, it's the same both times. We have found something that is true about adding numbers up.

However, mathematics doesn't stay like this for long, even though you might not have noticed it. Solving equations is a first example where this isn't quite the case anymore. If you're given an equation, say

$$3x + 4 = 10$$

you're saying: I *want* this to be true. What values of x make this true? In the end, a lot of advanced mathematics starts with a dream of something that you want to be true, and then you set to work building a world in which that dream is true. (I think this is how we make dreams come true in normal life too.)

Now what we're saying about infinity is "I want it to be the case that some infinities are bigger than others. What definition of 'bigger' will make this true?" Our prototype example is the apparent fact that there are "more" real numbers than natural numbers, although they're both infinite. We can use this as a definition of "bigger infinity" and see what happens. And what turns out to happen is rather interesting: we see how to build a hierarchy of bigger and bigger infinities.

Comparing Sets of Things

In the previous chapter we showed that the real numbers are uncountable by showing that any attempted pairing with the natural numbers was doomed to leave out at least one real number. We can use this to compare *any* two sets of things. It works for very small sets like these:

$$\begin{array}{ccc} 1 & \longrightarrow & 4 \\ 2 & \longrightarrow & 8 \\ 3 & \longrightarrow & 12 \\ & & 16 \end{array}$$

This example we saw before leaves out 16 on the right, but no matter what we try, we are doomed to leave out something on the right, even if it's not 16.

If every attempted pairing between sets *A* and *B* is doomed to leave something out on the right, then we say that the set on the right is "bigger."

The technical situation is slightly more complicated than this. The sentence above says that a surjective function from left to right is *impossible*, but technically we also have to say that an injective function from left to right is *possible*. This is rather subtle and related to the Axiom of Choice. In all the cases we'll think about, the injective function is very obvious; for example, from the natural numbers to the real numbers there's an obvious injective function that just pairs every natural number with itself within the real numbers.

In a way this is just like the very basic way in which small children compare how many things they have before they know the counting numbers. To see if they have enough cookies to give their friends, they might have to give them all out to find out, whereas someone who knows how to count will just count them. To see if we had the same number of *DWTS* celebrities as professionals, we just observed that they're all paired up so there must be the same number of each. We don't even need to count them because we didn't need to know how many there were. But even if we do count them, what we're really doing is matching the people in each group with the official number bag for that number. Say there are ten competitors. Then when you count them you're really pairing them up with the set

$$\{1, 2, 3, 4, 5, 6, 7, 8, 9, 10\}$$

If you start trying to pair them up with this set and discover you haven't used 9 and 10, you know that there are *fewer* than ten competitors. However, if you try to pair them up and discover that you'd have to use 9 and 10 twice, you know that there are *more* than ten competitors.

This analysis is a bit over the top for such small numbers, which is a typical problem with understanding mathematics. You have to start by understanding it in simple situations, but those are usually so simple that you don't need the complicated mathematics at all, so the whole thing seems pointless. But if you launch straight in trying to understand it in a really complicated situation, it's too hard.

Anyway, this way of pairing things up is the sort of thing a small child might do before they know how to count using numbers. They might have a train set and want to put a Lego person in each car. If they run out of Lego people and still have empty cars, it's obvious they don't have enough Lego people, even though they don't know how many Lego people they have, nor how many they need. This is how we have to deal with "more" and "fewer" when we're dealing with infinite sets, because, just like a child who can't count yet, we don't know "how many" things we have if we have infinitely many of them.

Let's try imagining we have infinite Lego people to put into an infinite Lego train, and we want to know if we have "more infinite" Lego people than train cars. Now, you might think that if we have empty cars left at the end, then obviously we didn't have enough Lego people. The first issue here is that we don't know what "the end" means; if we have to put all of our infinite people into cars (infinite or otherwise), it will take "forever" and there will be no end. We'll ignore that fact because we only have to come up with a scheme for doing it, we don't actually have to do it.

Here's a way we could put the people in cars and have an empty car left at the end. We could skip the first car, perhaps imagining that the Lego people are like me and don't like traveling in the front car in case there's a crash. We put the first person in the second car, the next person in the third car, and so on, and we fill up all the rest of the cars. In the end, we have one car empty, but this doesn't mean we didn't have enough people. We could move everyone up a car, and then all the cars would be full.

We could even have left infinitely many cars empty the first time round – maybe we only want to put the Lego people in even-numbered cars, so we put them into cars $2, 4, 6, \ldots$, and so on, just like in one of those Hilbert Hotel evacuation scenarios. Then all the odd-numbered cars will stay empty, which might make it look like we didn't have enough people. And yet, we do have enough people.

This is one of those weird and wonderful things that happens with infinity, which means we either have to give up, or we have to be more careful what we mean. Our infinite sets can be rearranged in weird ways, so that some arrangements will use up all the train cars and some arrangements won't, just like fitting extra people into Hilbert's Hotel. This doesn't happen if you have, say, ten people and ten train cars. If you want to put one person in each car, you can arrange them in plenty of different ways, but if you use up all the people, you'll always use up all the cars as well. If you have some other finite number of people and cars, and you use up all the people but still have empty cars, you know for sure that there were more cars than people.

With infinite people and infinite train cars, you can use up all the people and have cars left over, but you still don't know if you *might* be able to rearrange them to use up all the cars. If you really want to show you have more cars than people, you have to be sure there is *no possible* way of rearranging the people to use up all the cars. This seems hard, but it's what we've already done once with Cantor's diagonal argument. We showed then that there is no possible perfect pairing function between the real numbers and the natural numbers, because if there were, at least one real number would always get left out.

In mathematics we don't use the word "bigger" in this case because it's too ambiguous, and we're really talking about a very specific notion of "bigness" that we have defined rather carefully. The mathematical word for this is *cardinality*. The cardinality of a set of things is a measure of how many things there are in it. If the set only contains a finite number of things, then its cardinality is simply the number of things in it. If it contains an infinite number of things, it's more complicated. So far we're in a state a bit like a small child who hasn't learned to count – we know how to say when one set has the same cardinality as another, or greater cardinality than another, even though we don't know what its cardinality actually is. However, we can now build up infinite sets with greater and greater cardinality.

The Smallest Infinity

The smallest possible set is empty. So the smallest possible cardinality is 0. After that there are all the finite possibilities: a set with 1 object, a set with 2 objects, a set with n objects for any finite n.

It turns out that the natural numbers are the smallest possible *infinite set*. Remember, by "smallest" here, we are referring to our precise notion of size that we have defined using the method of pairing objects up. What would it mean to have a "smaller" infinite set of things than the natural numbers? It would mean there is some other infinite set satisfying this condition: if we try to make a pairing with the natural numbers, we are doomed to leave some natural numbers out.

Let's think about this in terms of evacuating people into Hilbert's Hotel again. It would mean that you have infinitely many people but when you evacuate them into the hotel, you are doomed to have wasted some rooms. But this can't be true: we could always re-evacuate them without wasting rooms by getting everyone to move into the new hotel in order, without skipping any rooms. That's a bit vague, but we could give precise instructions like this: "Count how many people there are with smaller room numbers than yours. Add one, and move into that room number."

* Whoever has the smallest room number will count zero people with smaller room numbers. So they move into room number $0 + 1 = 1$.
* Whoever has the next smallest room number will count one person with a smaller room number. So they move into room number $1 + 1 = 2$.

This is not quite a proof, but is the *idea* of a proof that every infinite subset of natural numbers can be paired up with the natural numbers. This means that every subset of natural numbers is either finite or has the same cardinality as the natural numbers. There is no infinity in between. So we have found the smallest possible infinity: it's the size of the natural numbers.

Remember when we were trying to see if infinity was a number, we kept running up against the problem of subtracting infinity from both sides of an equation. We kept finding that subtracting infinity from both sides of an equation led to a contradiction. We can now start to understand why. If you start with the infinite set of natural numbers, there are too many different ways in which you can remove an infinite subset. You could remove everything and be left with nothing. Or you could remove all the even numbers and be left with infinitely many odd numbers. Or you could remove everything bigger than 10 and be left with 10 numbers, or any other finite *n*. Subtracting infinity could produce *any* answer. We'll come back to this in the next chapter.

Now that we are constructing different sizes of infinity, we need to think of some better notation for infinity. In situations where we are being careful about different sizes of infinity, we write $\aleph_0$. This symbol is "aleph," the first letter of the Hebrew alphabet, and the subscript 0 is to indicate that this is the smallest possible infinity, and only the beginning of a hierarchy of infinities.

The Next Infinity

We have seen many infinities that turned out to be the same size as the natural numbers. There are ones that seem to be smaller but aren't, such as the even numbers or the odd numbers. We could be even more dramatic and think about only those numbers that are multiples of 100. Or those numbers that are multiples of a million. This is only a tiny proportion (one millionth) of the natural numbers, but it still gives us the same infinity as the cardinality of the natural numbers. This gives meaning to the idea that "infinity divided by a million is still infinity." We can also think about those numbers that are bigger than, say, 100. This still gives us the same size of infinity, and gives meaning to the idea that "infinity minus a hundred is infinity."

We have seen infinities that seem to be bigger but aren't, such as taking two copies of the natural numbers, one red and one blue. Or three copies, or a countably infinite number of copies. Or we could take the set of integers, or even the set of rational numbers. We have seen that all these infinite sets can still be paired up with the natural numbers, showing that these infinities are still $\aleph_0$.

So far we have seen only one set that is genuinely bigger: the set of real numbers. The question is: Is this the next infinity up? This is a very difficult issue addressed by the *Continuum Hypothesis*. The "continuum" here is the real numbers, because they are considered to fill up the entire number line "continuously" as opposed to the integers or the rational numbers, which still leave gaps between them. It's called a hypothesis because it hasn't been proved. It was suggested by Cantor in 1878, and proposes that the cardinality of the real numbers really is the next infinity up from the cardinality of the natural numbers. Another way of stating this is that there is no set whose size is somewhere between the size of the natural numbers and the size of the real numbers. It means that as soon as you manage to break free and become bigger than the natural numbers, you're immediately destined to be as big as the set of real numbers. Whether or not this is true cannot be proved. It is true in some worlds and false in some other worlds, so it depends what world you want to be in.

It has been proved that the Continuum Hypothesis *can't* be proved using a standard type of logic called Zermelo–Frankel logic. It has also been proved that it can't be disproved! This means that we could find a world in which it is true, and we could also find a world in which it isn't true. The result is thus *independent* of those rules of logic. The proof that it can't be disproved was by Kurt Gödel in 1940, and the proof that it can't be proved was by Paul Cohen in 1963. This was so important that Cohen was awarded the Fields Medal.

The next infinity up from $\aleph_0$ should be called $\aleph_1$, and so another way of expressing the Continuum Hypothesis is that the cardinality of the real numbers is $\aleph_1$. We can take this a bit further because in the previous chapter we worked out what the cardinality of the real numbers is, in relation to the cardinality of the natural numbers. The answer is that if the cardinality of the natural numbers is $\aleph_0$, then the number of real numbers comes out to be $2^{\aleph_0}$, using binary fractions.

You might notice that if we used decimal fractions the answer would come out to be $10^{\aleph_0}$. The binary version is more mathematically satisfying for several reasons. First of all, because 10 is a strangely arbitrary number to have floating around in our count of real numbers, and only comes from the fact that we have ten fingers and so favor base-10 decimal numbers. However, 2 is not so arbitrary – it's the smallest base possible.

Another reason that doing it with binary fractions is more satisfying is that $2^{\aleph_0}$ sort of *looks* smaller than $10^{\aleph_0}$, and part of what we're trying to do here is see if this is the next infinity up from $\aleph_0$. So it's more satisfying to be able to express it in the way that looks the smallest. It doesn't make any formal difference, because $2^{\aleph_0}$ is the same size of infinity as $10^{\aleph_0}$. In fact they're both the same as $1000000000000^{\aleph_0}$, but if I told you that this last thing was the next infinity up from $\aleph_0$, you might wonder whether the 1000000000000 part was important. So calling it $2^{\aleph_0}$ is more illuminating – it tells us that to make a bigger infinity, we have to raise something to the power of the previous infinity, even if it's as small as 2.

You might now remember that we ignored the part of the number before the decimal point (or binary point). This was a bit cavalier of me, but it doesn't really make any difference. We can attempt to fix it: the part of the number before the decimal place is the whole-number part, and we know how many whole numbers there are: it's $\aleph_0$. This means we need to multiply our count of the real numbers by $\aleph_0$ to allow for all the possible whole numbers at the beginning. However, multiplying the giant $2^{\aleph_0}$ by that countable infinity is not going to make the infinity any bigger. Secretly I knew that all along, which is why I wasn't very worried about the part before the decimal place.

A geometrical way to see this is that if we ignore the whole-number part, we are only counting the real numbers between 0 and 1. If we look at this portion of the line of real numbers, how many times do we have to stick this together to make the whole line? We have to stick one on for every natural number, so there are $\aleph_0$ copies of this portion. Moreover, we can restrict our attention to *any* portion of the real numbers, no matter how tiny, and still get the same total count of numbers in there. One way to see this is to make a bijection, that is, a precise pairing between the numbers in the tiny portion and the numbers from 0 to 1. For example, suppose we want to look at just the real numbers from 0 to $\frac{1}{100}$. We want to pair all these numbers up with the real numbers from 0 to 1. All we have to do is multiply them all by 100 and pair them up with the answer. This is like scaling the small piece of number line up to the bigger piece.

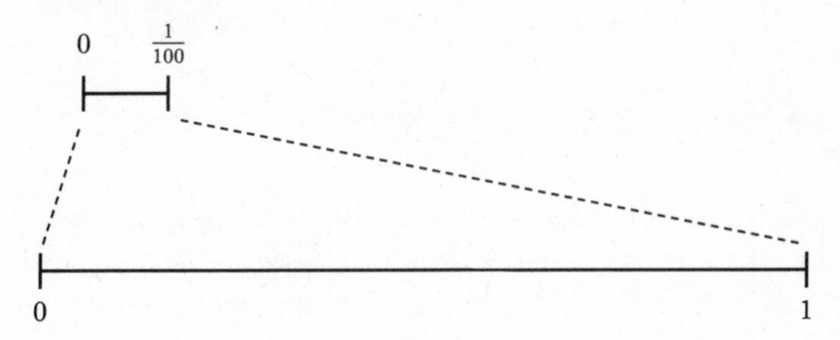

This argument would work for *any* size of interval, from any real number to any other real number. To scale the portion from 0 to 1 up to the whole of the reals, we have to do something more sneaky as we can't just "multiply everything by infinity." We could get all the positive real numbers like this: Start with your x taken from between 0 and 1. Then calculate $\frac{1}{x}$, which will give us something between 1 and infinity. Then subtract 1. This will give us a pairing of the real numbers between 0 and 1 with all the real numbers between 0 and infinity.

Now we know what the cardinality of the set of real numbers is, we can restate the Continuum Hypothesis as an equation:

$$\aleph_1 = 2^{\aleph_0}.$$

We have learned how to climb up one step of the staircase of infinities. If you're a mathematician type (just like a small child), you immediately want do it again and raise 2 to the power of this new infinity. The more general form of the Continuum Hypothesis says that, at every stage, this is the most minimal way to get a new bigger infinity. This would produce a hierarchy of infinities

$$\aleph_0, \aleph_1, \aleph_2, \aleph_3, \ldots$$

where in each case $\aleph_{n+1} = 2^{\aleph_n}$. As these are even more difficult versions of the Continuum Hypothesis, we can't prove that this is the minimal way to get a series of bigger infinities, but we can at least try to see that these infinities do get bigger.

Are These Infinities Really Bigger?

If $\aleph_0$ is "the size of the natural numbers," what could it possibly mean to raise 2 to the power of it? Usually we define 2^n to mean "2 multiplied by itself *n* times," but this won't work with infinity because you can't "multiply 2 by itself an infinite number of times."

The key here is to stop thinking about 2^n as something we do to numbers, but to remember we're thinking about cardinal numbers as the size of certain sets, the official number bags if you like. So *n* is really the size of a set of *n* things. We're going to rethink 2^n as the size of some set related to a set of *n* things. The idea is to do this in a way that makes sense for finite numbers, and then use it for infinite sets as well, because we know that $\aleph_0$ is nothing but the size of the set of natural numbers.

We're now going to see it's the same as choosing what shoes to take on vacation with you. When you're deciding this important matter, you could either just stare at all your shoes and pick the ones you want to take, or you could look at each pair in turn and decide: yes or no. So how many possible combinations of pairs of shoes are there that you could take with you? If you own two pairs of shoes in the world, say a pair of sneakers and a pair of sandals, then you have these choices:

* Take no shoes. Go barefoot.
* Just take the sandals.
* Just take the sneakers.
* Take both the sneakers and the sandals.

That's four possibilities. Another way of counting this would be to do the yes/no method, and we could draw a decision tree:

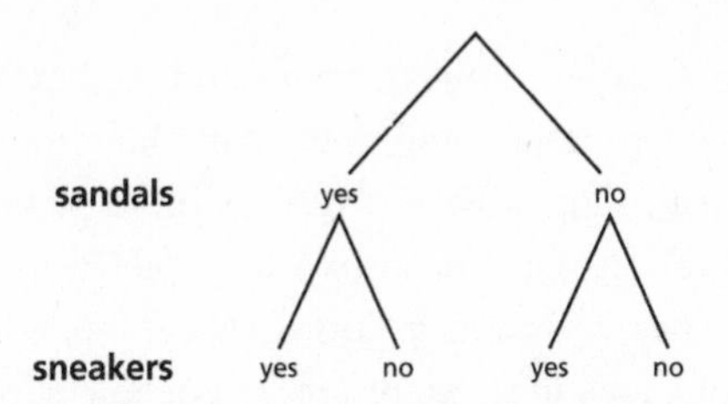

(Yes, if you actually drew this decision tree to help you decide on your two pairs of shoes, it would be absurd and everyone would laugh at you. This is sometimes the lot of a mathematician. I've been known to draw flow diagrams to optimize my kitchen time when I'm cooking for a big party, but I try to throw them away before anyone else sees them.)

You might recognize this tree from before. It's the same tree as when we were doing menu options and binary fractions. If we have three pairs of shoes, we get this tree:

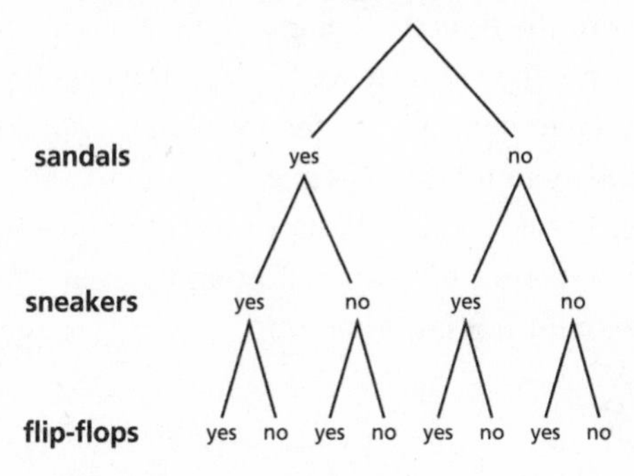

So now the total number of possibilities is $2 \times 2 \times 2 = 2^3 = 8$.

If we have n pairs of shoes, then the total number of possibilities is 2^n, just like the total number of binary fractions with n decimal places.

We now need to get a sensible interpretation of this when n is infinite. This is called a *generalization* and is an important part of mathematical thinking. You start with something fairly familiar and then try to expand the concept to make sense for something less familiar, without ruining the original concept. Previously we vaguely talked about "infinite trees" and took a leap of faith about it, but we're now going to do something more precise.

We have come up with 2^n while trying to pick which shoes to take on vacation. This means we were picking a *subset* of our total set of shoes. Our conclusion is that if our total collection has n pairs of shoes in it, there are 2^n possible subsets that we could take on vacation with us.

This is now a concept that works even if n is infinity – that is, if n is the version of infinity that we have come up with. We start with an infinite set, say the natural numbers. The natural numbers have cardinality $\aleph_0$, by our definition. Then we can consider the set of all possible subsets of natural numbers. Then we *define* $2^{\aleph_0}$ to be the cardinality of this set. We haven't calculated what size this is, but we have made sense of the notion of raising 2 to the power of infinity, in a way that is consistent with the notion for finite powers of 2.

We have no hope of ever writing down what all the subsets of the natural numbers are, but then we can't write down all the natural numbers in the first place either. In fact the situation is now worse, because the subsets of the natural numbers are uncountable – there's no way of putting them in a list, even in principle, without leaving some out.

If you remember that $2^{\aleph_0}$ is supposed to be the size of the real numbers, you might think it's a bit random to try and understand the "number of real numbers" via the "number of possible subsets of the natural numbers." In case it's bothering you, here's a way of seeing that those concepts are directly related. Remember that to pick a subset of the natural numbers you could either

- write a list of all the natural numbers you're including in your subset, or
- go down the list of all natural numbers and write "yes" or "no" next to them, to show if each one is in the subset or not.

If you use the second method, what you've effectively done is written a long string of "yes" and "no" that you could turn into a long string of 1's and 0's going on forever, and this is exactly what a binary fraction is.

It turns out we can use something just like Cantor's diagonal argument to prove that the collection of subsets of something is always bigger than the original collection. For finite numbers this is sort of obvious – no matter how many pairs of shoes you have, there must be more possible combinations you could take on vacation than there are pairs of shoes. For infinite sets this is a little trickier, but possible. And because this is how we're defining 2^n, this tells us that 2^n is always bigger than n. This means that our supposed hierarchy of infinities

$$\begin{aligned} &\aleph_0 \\ &\aleph_1 = 2^{\aleph_0} \\ &\aleph_2 = 2^{\aleph_1} = 2^{2^{\aleph_0}} \\ &\aleph_3 = 2^{\aleph_2} = 2^{2^{2^{\aleph_0}}} \\ &\vdots \end{aligned}$$

really does keep getting bigger. So no matter how big an infinity we've arrived at, we can always think about the collection of subsets of it and get a bigger one.

What this tells us is that when the children were arguing about how right they were and invoking infinity, the first way they could have made their infinity bigger was to say, "I'm right times two to the power of infinity!" And then:

"I'm right times two to the power of two to the power of infinity!"
"I'm right times two to the power of
two to the power of
two to the power of infinity!"
⋮

9 What Infinity Is

Sometimes when you're hiking in the mountains the path is very obvious. I'm not exactly a mountaineer, so the hiking I've done has mostly been on well-trodden paths, maybe even with signposts. However, there was one memorable time on a Duke of Edinburgh's expedition in France when we were almost at the campsite but confused by the last pathway. There were more paths than the map indicated. The compass bearing didn't seem to match any of them exactly. We were tired – I recall we had walked 42 kilometers that day, with our 20- or 30-pound rucksacks (tents were heavy in those days). We didn't want to go the wrong way and have to come back, so we compared various clues: which path seemed the most well trodden, which path seemed to match the compass bearing the best, which path seemed to lead somewhere that opened up. Finally we noticed an arrow in the dirt made from pebbles by some helpful person before us.

In this first part of the book we have hiked up various dead ends and retraced our steps, finally finding a promising path to infinity. We now have various clues about what infinity really is.

* Numbers can be measured as the size of an official number bag, or set. Infinity is no different.
* The smallest infinity is the size of the natural numbers.
* The infinity of real numbers is bigger than the infinity of natural numbers.
* Whatever infinity you are thinking about, if you take 2 to the power of that infinity, you get a bigger one.

The first point seems to indicate that, unlike what we saw in the first few chapters, infinity is in fact some kind of number, as long as we think about numbers in a certain way. The integers, rational numbers, and real numbers are extensions of the natural numbers, but in a way that didn't help us with infinity. We have seen that the "official number bag" way of thinking about numbers gives us an extension of the natural numbers that *does* help us with infinity. In fact, it gives us two different ways: the ordinal numbers and the cardinal numbers. We're now going to see what these are and see how they avoid the earlier problems we had when trying to define infinity as a number.

We have talked about cardinality as measuring the "size" of a number. The cardinal numbers are the official number bags that we use to measure the sizes of different numbers, especially bigger and bigger sizes of infinity. (The finite cardinal numbers are not very different in practice from the finite natural numbers.)

But there's another aspect of infinity we might care about. Thinking back to our first discussion of Hilbert's Hotel, we had a full hotel and just one new guest arriving. It seemed impossible to fit the extra guest in, but then we hit on the idea of moving everyone up one room. This is all very well but we completely neglected to worry about the hassle we were causing all those guests. If one guest arrives first and then an infinite busload of guests afterward, nobody has to move rooms in order for everyone to fit in: a hassle-free solution. But if the infinite busload arrives first and one extra guest afterward, all the earlier guests have to move rooms in order to fit everyone in: a necessarily "hassleful" solution.

This is about the order in which guests arrive. You might imagine that room 1 is the best, room 2 is the next best, and the rooms gradually get less good from there. In that case it would be unfair to put the last guest in the best room; you really want to assign rooms to guests in order of arrival.

This reminds me of the times I have stood in the queue at the Royal Albert Hall to go to the Proms, a series of concerts. When you arrive, you're given a raffle ticket telling you your place in the queue. I once arrived at 6 a.m. to queue to see the amazing Simón Bolivar Orchestra conducted by Gustavo Dudamel, playing Mahler's Second Symphony, one of my favorites. I was very excited to be number 6 in the queue, and after spending all day as number 6 I had become quite good friends with number 5. (Numbers 7 and 8 were my parents.) In this case it really matters what order everyone arrives in. I arrived that early because I really wanted to be at the very front, and after queueing for thirteen hours, I didn't really want anyone to push in front of me. Numbers 1, 2, and 3 had camped out overnight to get their places.

This is different from if you're hosting a party in a bar and the room has a maximum capacity of 100 people. You might give a raffle ticket to each person on the way in so you would know it was full after you'd handed out ticket number 100. You don't care in what order people arrived, you just need to know when you get to 100.

This is the difference between *cardinal* numbers and *ordinal* numbers. Cardinal numbers only measure the size of things, without really worrying about where those things all are in relation to one another. Whereas ordinal numbers take the ordering into account. This is like the question of the Hilbert's Hotel manager causing the guests hassle by moving them around to fit new guests in.

When we're only thinking about finite numbers, there's not a very noticeable difference between these two types of numbers. But when we're thinking about infinite numbers, they become very different. Ordinal numbers are much more subtle than cardinal numbers.

In the previous chapter we were talking about the sheer size of things, and we called it "cardinality." We were talking about cardinal numbers. One big difference between cardinals and ordinals is how many different types of infinity there are. We saw that the smallest infinity in terms of sheer size is the size of the set of natural numbers, and that the size of the set of real numbers might be the next infinity up. However, if we think about the order in which people arrive, and the hassle of moving rooms in Hilbert's Hotel, there are *definitely* ordinal infinities between those two infinities. If you care in what order people arrive, the situation becomes very different, and more strange.

The example of Hilbert's Hotel and one extra guest shows us that that "one plus infinity" fits into infinity with no hassle or unfairness, but "infinity plus one" fits into infinity only with hassle. We could write this fact like this:

$$\infty + 1 \neq 1 + \infty.$$

Now let's think about the example of two infinite busloads again. If you put the first busload into the hotel in order, you have a problem when the second busload arrives. You can fit them in by asking all the earlier guests to double their room number, leaving the odd-numbered rooms free. But then the new guests will be going into rooms ahead of the people who arrived earlier: unfair. What we have here is that "infinity times two fits into infinity, but it's a hassle."

Now suppose that instead of two groups of infinity turning up, you have infinity groups of two. So, perhaps, you have an infinity number of tandem bicycles arriving, each with two people on. A very realistic possibility, I'm sure you'll agree.

The first tandem arrives, and you put the front cyclist in room 1 and the back cyclist in room 2.

The next tandem arrives, and you put the front cyclist in room 3 and the back cyclist in room 4.

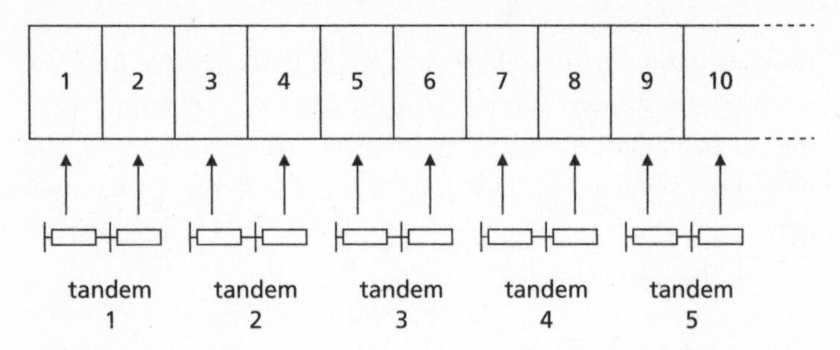

And so on. Every time another tandem arrives, you fill up the next two rooms. And the key here is that *nobody ever has to move rooms*. Where previously we had "infinity times two," as we had an infinity busload twice over, this time we have "two times infinity," as we have pairs of people but an infinite number of times. In normal arithmetic with finite numbers, we don't have to make the distinction between, say, "two times five" and "five times two" because we know they're the same. But here we find that "infinity times two" fits into the hotel with hassle, but "two times infinity" fits into the hotel with no hassle at all. That is, if we take hassle into account, we have the following fact:

$$\infty \times 2 \neq 2 \times \infty.$$

This is another hint that infinity does not behave like an ordinary number if we care about hassle.

You might be thinking that the notion of "hassle" is not a very mathematical notion, which is true. That's why we really think of it in terms of ordering. You can detect whether "hassle" has occurred by seeing whether the hotel guests are housed in rooms that match the order in which they arrived. If you don't care about the hassle you caused them, you can disregard the order in which they arrived – you just care about whether they fit in the hotel or not. This is called "cardinal arithmetic," because cardinality is just about how many things there are. If you do care about the order in which they arrived, this is called "ordinal arithmetic," and it's about what order things are in, not just how many there are. They're both the same when we're thinking about finite numbers, but get interestingly different when we think about infinity, giving us two different valid versions of infinity.

Infinite Queues

Now that we're thinking about the order in which people arrive, it's helpful to think about queues. Imagine you're in charge of handing out raffle tickets to mark people's places in a queue. This is going to be a miraculously big queue, a sort of "Hilbert's Queue" because there will be an infinite number of people in the queue. You have a book of raffle tickets with one ticket for every natural number: 1, 2, 3, 4, 5, 6, and so on. The people arrive, and you give each one a number in order of arrival.

Now suppose you've used up all those tickets, because an infinite number of people have arrived (just like in Hilbert's Hotel). If another person arrives now, you *could* just ask everyone to move back one place in the queue to free up ticket number one, and then you could give the new person that spare ticket. But then that last person will have jumped the entire infinite queue, which would not be fair at all. Instead, the only fair thing to do is start a new book of raffle tickets in a different color. Say the first book was red and the second book is blue. Then you have to remember that *all* the red tickets come before *all* the blue tickets.

You might wonder if there's some way of sorting this situation out so that you don't have to use a second book of tickets. The answer is that there is no way of doing it without some queue jumping happening. This tells us that according to *ordinal* numbers, "infinity plus 1" is actually bigger than infinity. When we are thinking about the infinity of natural numbers in this ordinal way instead of the cardinal (size) way, we often write it as ω. So what we have here is that $\omega + 1 > \omega$. On the other hand, $1 + \omega$ means that one person arrived first and then an infinite number of people arrived afterward. This way round you could quite happily use just one book of raffle tickets and never need the second book. This tells us that $1 + \omega = \omega$.

This is the beginning of the weird and wonderful arithmetic of ordinal numbers, where the rules we're used to with normal numbers all have to be reconsidered. The first rule that has changed is about the order of adding things up. With ordinary finite numbers, we know that order doesn't matter when we're doing addition. So

$$5 + 3 = 3 + 5$$

and more generally

$$a + b = b + a$$

for any real numbers a and b. (Remember the real numbers are all the decimal numbers, rational and irrational.) We've already seen an example involving infinite ordinal numbers where this fails, because we've seen that

$$1 + \omega = \omega$$

but

$$\omega + 1 \neq \omega$$

so

$$1 + \omega \neq \omega + 1.$$

The property $a + b = b + a$, if it is true, is called *commutativity of addition*, because the a and the b can "commute" past each other. It is still true for finite numbers even if we are thinking about this new ordinal way of doing things. Let's try it with 5 and 3.

Suppose you're in charge of that queue again, and five people turn up. You will give them raffle tickets 1, 2, 3, 4, 5. If three more people turn up, you will give them 6, 7, 8. This shows that $5 + 3 = 8$ with ordinal numbers.

Now, if three people arrive first, you give them tickets 1, 2, 3. If five more people turn up, you give them tickets 4, 5, 6, 7, 8. The answer is still 8. You didn't *have* to give them those tickets in that order, but the fact is that in both cases you *can* give them tickets 1–8 without disrupting the order in the queue. So the answer is the same, no matter what order you do it.

If you were feeling a bit strange, you could have handed out the tickets 2, 4, 6, 8, 10, 12, 14, 16 instead, or even 1, 5, 9, 14, 18, 100, 200, 378. This is valid, and would mean that everyone was still in the order they arrived; you'd just have to remember not to give out the tickets you skipped when more people arrived.

In the case of 1 and ω, there is no way of handing out raffle tickets so that one person followed by ω people uses the same tickets as ω people followed by one person without disrupting the order of the queue. You might notice that this is not a mathematical argument – it's just a plausible thought. The genuinely mathematical way of showing that $1 + \omega$ is not the same as $\omega + 1$ is to think about who is the last person to arrive in the queue. In the first case we can't tell who it is, because the ω people who arrive go on forever. Whereas in the second case we know exactly who is the last person to arrive – it's the one who arrived *after* the infinite number of people who came first. Any queue with an "identifiable last person" isn't the same as a queue with no identifiable last person.

In mathematics, the queues with no last person are called "limit ordinals," because you've used up a book of raffle tickets all the way to its limit and you can't continue without getting hold of another book of raffle tickets.

Note that there are plenty of queues of different lengths that all have an identifiable last person – for example, all finite queues. Soon we'll see that there are also plenty of queues of different lengths that *don't* have an identifiable last person.

Multiple Infinite Queues

The next thing to try after addition is usually multiplication. This is because multiplication is usually built up from doing addition repeatedly. With ordinary numbers, we know that the order of multiplication doesn't matter, so

$$5 \times 3 = 3 \times 5$$

and more generally

$$a \times b = b \times a$$

for any real numbers a and b. This is called *commutativity of multiplication*. We have already had a hint that this doesn't work for infinite numbers, when we thought about the difference between an infinite number of tandems arriving at Hilbert's Hotel compared with two infinite busloads of people.

At this point we have to be careful about what multiplication really means, because there are different ways of building it from repeated addition. We don't have to worry about it with ordinary numbers because the different ways give us the same answer. You might have entirely forgotten that these two different ways of thinking about it are even there. The two different ways are this: Is 5×3 supposed to be "five lots of three" or "three lots of five"? Is it five bags of cookies where each bag has three cookies, or is it three bags of cookies where each has five cookies? Is it $3 + 3 + 3 + 3 + 3$ or $5 + 5 + 5$? With ordinary numbers we don't worry because we know it's the same answer. However, if you help small children with multiplication, it will take some time to convince them of this, and usually involves lining some objects up in a rectangular grid pattern.

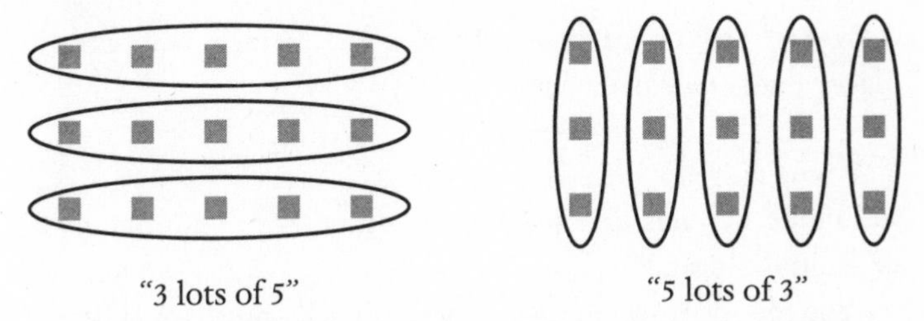

"3 lots of 5" "5 lots of 3"

In the case of infinite ordinals, it *is* going to make a difference, so we have to make a decision about what a times b means. The usual decision in mathematics is that $a \times b$ means "a added to itself b times." This is sort of consistent with a^b meaning "a multiplied by itself b times."

So for 5×3 we have to think about 5 people arriving in the queue, followed by another 5, and then another 5. You will hand out raffle tickets 1–15. For 3×5 we have to think about 3 people arriving in the queue, and then another 3, and then another 3, and another 3, and another 3. You will still be able to hand out raffle tickets 1–15.

However, now let's think about the busloads and tandems again. If ω people turn up, followed by another ω people, you will have to use up a whole book of, say, red tickets, followed by a whole book of blue tickets.

On the other hand, if the infinite tandems turn up, you would start by handing out tickets 1 and 2 to the first tandem, then 3 and 4, then 5 and 6, and so on. You could keep going like this and never have to break open a second book of raffle tickets.

In the first case we have done "ω added to itself 2 times," which we're calling $\omega \times 2$. The answer is two whole books of raffle tickets, which is the same as $\omega + \omega$. In the second case we have done "2 added to itself ω times," which in our definition of multiplication is $2 \times \omega$. This time the answer is just one book of raffle tickets, that is, ω.

$$\omega \times 2 = \omega + \omega$$
$$2 \times \omega = \omega$$

Now $\omega + \omega$ is even bigger than $\omega + 1$, which was already bigger than ω. This means that

$$2 \times \omega \neq \omega \times 2$$

so infinite ordinals do not have the property of commutativity of multiplication.

You might now wonder if we can tell these are different using the same method as last time, by seeing who is last in the queue again. However, this time it won't work as neither of these queues has a last person. We have to use a slightly different approach: we can imagine asking everyone to see who is directly in front of them in the queue. For the queue with ω people, the only person who doesn't know who's in front of them is the person with raffle ticket number 1, because there's nobody in front of them. Whereas for the queue with $\omega + \omega$ people, there are *two* people who don't know who's in front of them:

* the person with red number 1, because there's nobody in front of them, and
* the person with blue number 1, because all the people with red tickets are in front of them but there is no "last" red person, so there is no identifiable person directly in front.

We can use this method to show that every time we add another ω we'll get a new ordinal number, because each time there will be a new number 1 person with a new color of ticket who won't be able to tell who is directly in front of them.

Infinite Subtraction

We can now think about subtraction, the thing that was causing us problems right at the beginning when we were trying to see if infinity could be an ordinary number. When we defined subtraction for integers, we did it using the notion of additive inverses, that is, the negative numbers. However, with the ordinal numbers we don't have negative numbers. Who would be the minus-first person in a queue? You can imagine a small child trying to claim this place in an attempt to beat the person who is first in the queue, but it doesn't really work like that. Not unless you can then add that person to another queue to eliminate a person from that queue. (Maybe the small child would like to do that too, and would at this point get cross with the entirety of mathematics, like when you try and get them to do arithmetic by thinking about numbers of cookies but say they're not allowed to eat them.)

Instead we have to go back to the way you do subtraction when you're little and you don't know how to count backward yet. Let's think about $5 - 3$. We all know the answer is 2, because we can count backward from 5, or because we can put up five fingers and then put two of them down and count the rest. Or just because we know. However, if you only know how to count up and not down, you can do this by starting at 3 and seeing how many steps it takes to count up to 5. When children do this, they are answering the question "What do I have to add to 3 to make 5?" We could get a bit fancy and write that as an equation:

$$3 + x = 5.$$

There's another related question: "Is there a number such that if I add 3 to it, I'll get 5?" Note the crucially different wording; this is expressed in the equation

$$x + 3 = 5.$$

Now in normal numbers, these are the same equation, because $3 + x = x + 3$, but we've seen that for infinite ordinals, addition is *not* commutative, so with other numbers in place of 3 and 5, these could be different equations. We need to think about these two situations separately when we're thinking about infinite ordinals.

For example, what about this: What do I need to add to 1 to get to ω?

$$1 + x = \omega.$$

This would be solved by putting $x = \omega$ because we know that

$$1 + \omega = \omega.$$

If we try it the other way round, things go wrong. Is there a number such that if I add 1 to it I'll get ω? This is:

$$x + 1 = \omega.$$

We can't put $x = \omega$ anymore, because

$$\omega + 1 \neq \omega$$

as we saw earlier. In fact, there are no possible solutions, because no matter what we put for x, $x + 1$ is going to have a "last person in the queue" (the 1 at the end), whereas ω will not have a "last person in the queue." This means that we can't necessarily solve equations like

$$x + a = b,$$

which means we can't necessarily do subtraction. Let's see why.

We have to remember that adding things on the left and right are different with infinity, so subtracting on the left and the right will probably also be different. Subtraction is "undoing addition." If we undo addition on the left of the number, we are removing people from the front of the queue. If we undo addition on the right of the number, we are removing people from the back of the queue, and here's the problem: if we have no identifiable last person in the queue, we can't remove the last people in the queue because we can't find them. This is why we can't solve

$$x + 1 = \omega$$

because to find x we have to undo the $+1$ on the right of ω. Whereas we can solve

$$1 + x = \omega$$

because this one involves undoing the $+1$ on the left of ω, the front of the queue.

This is exactly the problem we kept running up against before when showing that infinity couldn't be any of the earlier types of numbers. We said we wanted

$$1 + \infty = \infty$$

but that would mean we could "subtract ∞" from both sides to give

$$1 = 0.$$

We've finally found a way of defining infinity that doesn't cause this problem. We have

$$1 + \omega = \omega$$

but we *can't* subtract ω from the right of both sides, as we can only subtract from the left. If we try subtracting ω people from the front of the $1 + \omega$ queue, there will be nobody left, just like if we subtract them from the front of the ω queue. So subtracting ω from the left on both sides of this equation just gives

$$0 = 0$$

as expected. We finally have a logically valid way of defining infinity that accounts for the strange behavior with subtraction, so we finally have an answer to the argument between my little nephew and his friend, about whether infinity is a number or not.

Infinity isn't a natural number,
an integer,
a rational number,
or a real number.
Infinity is a cardinal number and an ordinal number.

Cardinal and ordinal numbers do not have to obey all the rules that the earlier types of number obey. This is why infinity works here.

We also have some new answers to the children arguing about being right "times infinity." If they are arguing with ordinal numbers, then they have a lot more scope for subtly different levels of infinity than if they are arguing with cardinal numbers. With cardinal numbers you have to leap up to "2 to the power of infinity" to get a bigger infinity with which to be right. But with ordinal numbers you only have to add 1: you just have to be careful to add it on after your infinity, not before it.

My favorite example of this is from *The Taming of the Shrew*:

If this be not that you look for, I have no more to say,
But bid Bianca farewell for ever and a day.

If forever is ω, then forever and a day is $\omega + 1$. Shakespeare knew that $\omega + 1$ is more than ω. At least, that's what I like to think.

part two

THE SIGHTS

10 Where Is Infinity?

I wish I could teleport. I would love to be able to think of a place I'd like to be and instantly be there. I would still go on normal journeys if they're beautiful, but if I just need to be somewhere to accomplish something or visit a long-lost friend, it would save a lot of time and effort to be able to teleport.

One of the things I enjoy the most about the abstract world is that everything does happen as soon as you think of it. If I want to explore a different abstract world, I *can* just think of it and instantly be there. If I want to play with a new abstract toy, I can do it as soon as I've thought of it. As soon as you have an idea, that's it – the idea exists. If only I could make my dinner exist just by getting an idea in my head of what I want to eat.

Of course, we could have long arguments about what it means for an abstract thing to "exist," but for me it means that I can play with it. This is how abstract mathematical research often works. You have an idea for a new mathematical concept, and you can immediately start playing with it in your head – building things out of it, seeing how it interacts with other things. It might cause a contradiction and make everything implode, but the idea still exists and you can still play with it. It's not like having an idea for a new type of car or medicine where you then have to work out how to turn it into "reality" – get equipment and materials to manufacture it, get funding to obtain the equipment and materials, and so on.

For example, the idea of numbers exists because we've thought of it. Does that mean numbers exist? It depends who you ask. Philosophers get into long arguments about this. Personally I am happy with the thought that a number *is* just an idea. I don't worry about whether or not numbers exist any more than I worry whether or not I exist. Whether I "exist" or not, I still get on with my life. Whether numbers "exist" or not, I still get on with mathematics. Then again I might be unusual in my attitude to existence: I'm the only person I know who thinks Father Christmas exists on the grounds that Father Christmas is the idea that causes presents to be given at Christmas. Some people would say that this only means the *idea* of Father Christmas exists, but I am quite happy to say that Father Christmas is an abstract idea, just like numbers are an abstract idea, and be done with it.

In this sense, I am happy to say infinity exists, on the grounds that it exists as an abstract idea. But we can still ask ourselves whether infinity exists at any less abstract level than this. Does any infinite number of things exist in "real life"? I always hesitate over the term "real life" for several reasons. First, because abstract things aren't necessarily less real than "real" things. Tiredness is abstract but feels very real to me, whereas the bottom of the Pacific Ocean is real but feels very abstract to me as I will never touch, see, or experience it. The other reason I'm wary of the idea of "real life" is that too many supposedly real-life mathematics questions are completely implausible, involving things like your wild horses escaping, or going out to buy seventy-five watermelons.

It is questionable whether an infinite number of "real things" is present in the universe. There is certainly an extremely large number of molecules or atoms or electrons, but as we've seen, there's a big difference between an "extremely large number" and "infinity." The universe itself seems infinite to our small finite brains, but might be finite (it is currently unknown).

There certainly are infinite quantities of *abstract* objects. Take numbers, for example. We know that the natural numbers definitely have no bound to them, unlike the universe, where we can't tell. Each individual natural number is finite, but they get bigger and bigger forever. In the sense that numbers exist at all, there is an infinite quantity of them. In the next chapter we're going to look at some other things that keep growing so much that we can't put any boundaries on them. Like the natural numbers, they are finite at any given moment but grow so much that it is useful to think of them as "approaching infinity," and it is possible to make sense of this. I have a friend who adopted a Great Dane puppy and he grew so implausibly fast that at one point it seemed as if he might be approaching infinity.

In Chapters 12 and 13 we'll look at ways in which our brains have an unbounded capacity even if our brains aren't themselves infinite. There is no limit to the gradations of subtlety we are capable of, and this leads us to abstract higher-dimensional spaces in Chapter 12. These are different from physical spaces, as they are realms of thought rather than realms of physical existence. We'll see that we can give meaning to the informal idea of a book's storyline being "a bit one-dimensional," and that if we think about dimensions in that sense, our lives are very multidimensional. The number of dimensions can't really be bounded, so it again approaches infinity. In Chapter 13 we'll think about a different type of dimension, a type that comes from category theory, a field of abstract mathematics that studies relationships between things. Here we think about the nature of relationships between things, but then we think about the relationships between those relationships, and the relationships between those relationships, and so on "forever," leading us to infinite-dimensional categories. Both of these types of dimension give us more subtle and expressive ways of thinking about the world around us.

We'll then move on to a different way in which things can seem to be infinite. Rather than something being so big as to be unbounded, we might want to understand something in tiny portions. This is where we realize that if we think about infinitely small portions of things, everything is divided into infinitely many of those infinitely small portions. In Chapter 14 we'll see how thinking of infinitely small things opens up some strange paradoxes that took thousands of years to resolve. In Chapter 15 we'll see that in resolving those issues, mathematicians realized they didn't really know what real numbers were, at least not well enough to make formal arguments about them as precisely as they needed to. In Chapter 16 we'll look at some weird things that arise from this new understanding of infinite and infinitely small things. Finally, in Chapter 17 we'll investigate ways in which infinity implicitly arises in our lives, via loops that we can go round forever. But what does "forever" mean?

The Abstract Version of Forever

One way in which infinity comes up in life is when we talk about things "going on forever." We have decimal numbers that "go on forever" and the natural numbers "go on forever." In *real* real life nothing can go on forever, but in mathematics we have an abstract version of this that enables us to gain access to forever in an instant. This is another joy of abstraction for me.

Right at the beginning of the book I mentioned my favorite computer program, which prints infinite lines of HELLO on the screen. It only needs two lines of code though: one line to start us off, and one line to say "do this again." This is how we produce forever abstractly. It's just like the way we saw of constructing the natural numbers where we start with 1 and keep on adding 1 to it "forever." We do this *abstractly* rather than actually going through the motion of adding 1 repeatedly. If we went through with the physical motion of adding 1, and then another 1, and then another 1, it would take "forever," a longer time than we're alive. However, if we do it abstractly, then we get all the natural numbers all at once. It's like doing it in *principle* rather than in practice. One of the fun things about the abstract world is that doing things in principle and doing them in practice turns out to be the same. There really isn't much difference, unlike in that wry saying "In theory, theory and practice are the same, but in practice . . ."

This way of making all the natural numbers is the basis of something called the *principle of mathematical induction*. It essentially says that you can think about all the natural numbers at once, just by thinking about the number 1 and by thinking about adding 1 to everything (in principle).

What it actually says is that if you want to show something is true for every natural number, you don't have to go round proving it for every natural number. You just have to prove:

- a starting point: that it's true for the number 1, and
- a way of going up 1: that *if* it's true for n, *then* it's true for $n + 1$.

Once you've done that, then suddenly, just like that, you know it's true for all the natural numbers.

This is like a small child who discovers that once they've learned how to climb up one step, all they need is for some adult to plonk them at the bottom of the stairs and they can then climb up stairs as far as they want (until they are removed). In principle they can climb up forever, although in practice, even if they're not removed, they will get hungry and tired. Mathematical objects do not get hungry or tired. Whenever we are in a mathematical situation where we could in principle do something one more time, that means there is no end to how many times we could do it, and this is what it means for something to go on "forever" in mathematics. The numbers go on forever, a sequence of numbers can go on forever, a decimal number can go on forever.

The fact that we have an infinite quantity of natural numbers gives us access to all sorts of other abstract infinite things as well, in both directions: larger and smaller. We have already seen how we can start with the infinite set of natural numbers and produce infinite sets that are progressively bigger and bigger. But where, if anywhere, can we find such infinitely large sets of things in the world around us? One way is if we consider things that are infinitely small. This is the basis of calculus, the field of mathematics that is one of the most ubiquitous in its applications. Calculus is the study of things that are changing. It is difficult to make theories about things that are always changing, and calculus accomplishes it by looking at infinitely small portions, and sticking together infinitely many of these infinitely small portions. These infinitely small things are everywhere around us, making up infinite sets of things that we actually experience every day without needing to be aware of it. In this second part of the book we'll look at the ways infinity pops up around us whether we're aware of it or not.

One of the strange features of mathematics is that it is there whether or not we see or understand it, just like when we're on a train the scenery is there whether or not we look out of the window or know what it is that is going past. Understanding the mathematics that is there can enable people to build better systems and solve more complicated problems. But it can also shed light on how we interact with the world around us, and on how our own minds work. Illumination is a more subtle and less dramatic result than solving a specific problem or building a specific technology, but to me it is very fundamental and far-reaching in its quiet importance.

11 Things That Are Nearly Infinity

I am almost certain that I will never go to the top of Mount Everest. I will optimistically leave open the possibility of teleportation, but apart from that, I am sure I will never go. I will also almost certainly never go to the South Pole. I don't know anyone who's climbed Mount Everest, but I do know an astrophysicist who works at the South Pole. I know that the South Pole is difficult to get to, even by plane, but is still only finitely far away. I know that Mount Everest is only finitely high. But to me they both might as well be infinitely far away because I will never go there.

Infinity exists, but can we ever get there? Can we ever do infinitely many things, perhaps if they're infinitely small enough? Before we really look at how we can make sense of this, we're going to think about things that seem to get so big they might almost be infinite, and times we seem to be doing something almost infinitely.

There's an old conundrum about rice on a chessboard. The story is that a man asks for one grain of rice on the first square of a chessboard, double that on the second square, double that on the third square, and so on for every square until the chessboard is full. The question is: How much rice will he end up getting? The short answer is: rather a lot. But exactly how much?

It's not a difficult question in principle, because you just have to keep multiplying by 2 and adding all the answers together until you've done all 64 squares. However, if you try this, you will discover that the numbers get big awfully quickly, much bigger than your calculator or even your computer can handle in normal settings (unless you have some special computational tools on there). There is a trick for speeding up the calculation, but you still end up having to deal with a very, very large number: 18,446,744,073,709,551,615 grains of rice.

Of course, we don't usually measure rice in grains, except in absurd-sounding math questions. (I first heard about this question in a math lesson and tried to work out the answer by hand. I got it wrong.) So how much rice is this in practical terms? I just tried to weigh 1 g of rice and then count the grains, and it seemed to be about 50. So we could do this rough approximation:

		1 g	=	50 grains of rice
1 bowl	=	100 g	=	5000 grains
1 person	=	4 bowls of rice per day	=	20,000 grains
the world	=	7 billion people	=	140,000,000,000,000 grains
a year	=	about 500 days	=	70,000,000,000,000,000 grains

This has 16 zeros on the end. The number of grains we had was 18,446,744,073,709,551,615, which is approximately 2 with 19 zeros on the end: that's 3 more zeros, a factor of about 1000. So it looks like we could feed the world's population for around 1000 years. (Not taking into account the fact that, the way we're going at the moment, the world's population is growing a lot each year.)

My calculation was extremely crude, but gives the general idea: just by doing some innocuous doubling of quantities as you move around a chessboard, you quickly get to an impossible quantity of rice, more rice than currently exists in the world.

Puff Pastry

Puff pastry relies on the same principle, that repeated multiplication makes things grow extremely fast. Puff pastry has an apparently miraculous number of tiny layers in it, and they are created by folding the dough in three just six times. The dough has a thick layer of butter sandwiched in it to start, at just the right consistency so that when you roll it, the butter flattens neatly inside its sandwich. Then you fold it in three, making six layers, and chill it so that the layers stay firm and don't start melding into each other. Then you roll it out, fold it in three and chill it again. You do this six times. Repeatedly multiplying by three makes the number of layers grow very fast, and then when you bake the pastry, the thin layers of butter melt, the liquid part of the butter evaporates and creates steam, and this pushes the layers apart so you can watch the pastry physically grow in the oven, not just the numbers abstractly growing.

This is my favorite demonstration of exponential growth. Informally, people say things are growing exponentially just to mean they're growing a lot, which is sort of true, but the formal mathematical meaning is that it's growing at the same proportional rate all the time. If I folded the puff pastry in three the first time, and then four, and then five, and then six, the number of layers would grow even faster, but it wouldn't be exponential as the rate of multiplication is changing.

I love the fact that exponential growth directly translates into deliciousness in puff pastry. The multiple layers of pastry are not just dramatic and beautiful, but so thin that they melt away delicately in one's mouth. Puff pastry has a reputation of being difficult to make, but I think the genius of the method is that the use of exponentials actually makes it rather easy to create those incredibly thin layers of pastry. After all, it would be very difficult to roll out such thin layers individually. And the whole point of mathematics should be to make difficult things easier.

Unfortunately it often comes across as a way of creating difficult things out of nowhere.

The iPod Shuffle

I remember when the iPod Shuffle first came out, I saw this big advertisement on the Tube with the slogan "240 songs. A million different ways." The idea was to impress people with the fact that with a mere 240 songs, you could have a million different ways of playing them if you played them in randomly different orders instead of just going through the library from start to finish.

In fact, this is a vast underestimate. I sat on that train calculating, just for amusement value, how many songs you would really need in order to achieve a million different ways of playing them.

If you have two songs, there are just two ways of playing them. Either you start with one and then play the other, or you do it the other way round. Let's say you have three songs. For the first song, you have a choice of three you can play. For the second, you only have a choice of two remaining songs, and for the third, there's no choice anymore. (We are assuming that you don't want to listen to the same song again, although in reality I often listen to the same song on repeat for hours.) We can again draw this in a tree diagram, except this time there will be fewer and fewer branchings at each level as we gradually run out of songs, assuming you don't like replaying them.

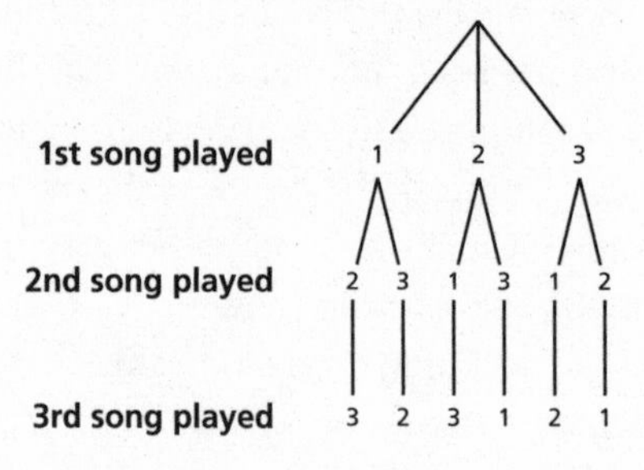

So we see that there are 6 possible orders here, each one being read from the tree by following a path down it. We could have calculated it as $3 \times 2 \times 1$ using the number of possible choices at each step.

Now if we have four songs we get this tree:

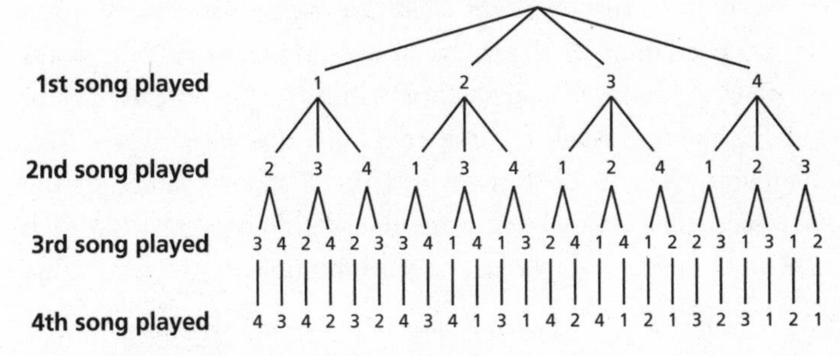

or by calculating:

* 4 possibilities for the first song,
* 3 possibilities for the second song,
* 2 possibilities for the third song, and
* 1 possibility for the last song.

So the total number of ways is $4 \times 3 \times 2 \times 1$. (We don't really need to write the 1 at the end as multiplying by 1 doesn't change anything.)

This is called a factorial in mathematics, and "four factorial" is written as "4!." In general, n factorial is

$$n \times (n-1) \times (n-2) \times \cdots \times 4 \times 3 \times 2 \times 1.$$

We can also define it by induction, a loop a bit like my infamous `HELLO` program:

* $1! = 1$
* $(n+1)! = (n+1) \times n!$

So if we have n songs, the total number of orders we could play them in is $n!$, because we have n possibilities for the first song, $n-1$ possibilities for the second song, $n-2$ possibilities for the third song, and all the way down to two possibilities for the second to last song, and only one possibility for the last song.

Now, the question I was trying to ask is: How many songs do you need in order to achieve at least a million different ways of playing them? Stated mathematically, this means we're looking for the smallest value of n such that $n!$ is bigger than a million. We can just write out the first few values of the factorials until we make it, remembering that to get from each row to the next row, we just have to multiply by the next value of n.

$$\begin{aligned}
1! &= 1\\
2! &= 2\\
3! &= 3 \times 2 = 6\\
4! &= 4 \times 6 = 24\\
5! &= 5 \times 24 = 120\\
6! &= 6 \times 120 = 720\\
7! &= 7 \times 720 = 5040\\
8! &= 8 \times 5040 = 40{,}320\\
9! &= 9 \times 40{,}320 = 362{,}880\\
10! &= 10 \times 362{,}880 = 3{,}628{,}800
\end{aligned}$$

Bingo: we have burst through a million. We only need 10 songs, and we'd already have over 3 million ways to play them.

The other question we can now ask is: With all 200 songs, how many ways to play them do we really have? We have to calculate

$$240 \times 239 \times 238 \times \cdots \times 3 \times 2 \times 1.$$

This is going to be hopeless without special equipment, as the numbers are much too big. A quick experiment with an ordinary computer spreadsheet finds that it can only manage up to 17! before resorting to an approximation, and we have

$$17! = 355{,}687{,}428{,}096{,}000.$$

I'm tickled to find that my phone's calculator can go further

$$18! = 6{,}402{,}373{,}705{,}728{,}000.$$

My phone's calculator then resorts to approximations, but completely gives up after telling me that

$$103! \simeq 9.9 \times 10^{163},$$

and for any larger attempts it just gives me an error. My computer's spreadsheet keeps plowing on with approximations until 170

$$170! \simeq 7.3 \times 10^{306}$$

and then it, too, gives up. This is because of the way that my (old, feeble) computer uses its memory to hold large numbers. 171! turns out to be the first factorial that is too large to hold.

Computational statistician Rick Wicklin has written a program to compute large factorials, and as it happens he has published the value of 200! on his blog. He gets round the problem by making the computer store the large numbers as a very long string of individual digits rather than as a large number. Then he gets the computer to do the multiplication just like you would by hand, place by place, carrying over the digits to the next column where necessary. The answer he gets for 200! is this extraordinary number:

7886578673647905035523632139321850622951359776871732
6329474253324435944996340334292030428401198462390417
7212138919638830257642790242637105061926624952829931
1134628572707633172373969889439224456214516642402540
3329186413122742829485327752424240757390324032125740
5579568660226031904170324062351700858796178922222789
62370389737472000000000000000000000000000000000000
00000000000.

I believe this is the biggest number I have ever physically looked at. It has 375 digits, making it very, very, very many orders of magnitude bigger than the "1 million ways" stated in the advert, and still not as big as 240!

While playing around with these absurdly large numbers, I thought I might as well see how many songs you'd need on your iPod to get more ways of playing them than the number of grains of rice on the chessboard. The answer: 21. What this tells us is that while doubling numbers makes them get big very fast, taking factorials makes them get big *even faster*.

> You might be wondering if it's a weird fluke that there are so many 0's on the end of that number: 49 of them in fact. Large factorials are destined to have a large number of 0's on the end, and we can work out how many there will be even without knowing the answer to the factorial. Each 0 comes from a factor of 10 in the final answer, and each 10 comes from a factor of 2 and a factor of 5 in the individual components of the factorial. There are way more factors of 2 hanging around than 5's, so we just need to count how many multiples of 5 there are between 1 and the number whose factorial we're taking, remembering that some numbers have multiple factors of 5 so will contribute multiple 0's.

How Fast Do You Grow?

Children grow very quickly when they're little, and it seems even quicker because they're so small. Even if a child grows at around, say, 10 centimeters a year for the first few years, that amount of growth is much more dramatic for a baby as it's a much bigger proportion of the baby's total size. Of course, children grow at different rates, and some children who are shorter when they're younger can have a growth spurt later on and start overtaking everyone.

In mathematics we also think about how quickly things grow, and whether some things grow more quickly than others. In the case of the rice, for example, we're thinking about 2^n where n is getting steadily bigger. In the case of the songs, we're thinking about $n!$ where n is getting steadily bigger. In both cases the numbers get unimaginably big very quickly, although they're always still finite. Mathematically we say that "2^n tends to infinity as n tends to infinity," carefully avoiding ever claiming that something *is* infinity. The factorial $n!$ also tends to infinity as n tends to infinity. But we also sense that $n!$ goes "faster" than 2^n. What does that mean?

One way we can make sense of it is to pit one against the other by putting them in a fraction

$$\frac{n!}{2^n}$$

and seeing who "wins" as n gets bigger. If the fraction keeps getting bigger, that means $n!$ is winning. If the fraction keeps getting smaller, then 2^n is winning. If the fraction stabilizes, this will show that it's a draw. Now we can do a bit of a clever trick again. We can write out that fraction as

$$\frac{n \times (n-1) \times (n-2) \times (n-3) \times \cdots \times 4 \times 3 \times 2 \times 1}{2 \times 2 \times 2 \times 2 \times \cdots \times 2 \times 2 \times 2 \times 2}$$

and then we can split it into individual fractions like this

$$\frac{n}{2} \times \frac{n-1}{2} \times \frac{n-2}{2} \times \frac{n-3}{2} \times \cdots \times \frac{4}{2} \times \frac{3}{2} \times \frac{2}{2} \times \frac{1}{2}.$$

Now we can see that this is a product of a whole load of fractions that are *almost all* top-heavy (apart from the one at the very end, $\frac{1}{2}$). Moreover, as n gets bigger, this product of fractions is going to contain progressively more terms, and the new terms are going to be *even more* top-heavy than before because while the top (numerator) gets bigger, the bottom (denominator) stays as 2 all the time. The end result is that the top is *emphatically* winning over the bottom.

There is a sort of hierarchy of how quickly things tend to infinity. It's a bit different from the previous hierarchy of infinities that we had, but has a similar idea.

We've just seen that $n!$ grows faster than 2^n. Well 2^n in turn grows faster than n^2. In fact 2^n grows faster than n^3 or n^4 or n to the power of *anything*, even $n^{10000000000000000000}$. This last thing seems huge, and it will be bigger than 2^n at the beginning (except when $n = 1$), and yet 2^n will eventually overtake. We can try this for something less absurd, say n^{100}. My computer tells me that n^{100} wins until $n = 125$, and then 2^n overtakes.

We are only thinking about *positive* powers of n because negative powers don't actually grow at all – they get smaller as n gets bigger.

Something that grows more slowly than all possible powers of n is $\log n$. You might remember that logarithms are the "opposite" of exponentials. If we're doing it in base 10, then $\log n$ is "the power you would have to raise 10 to to get n." So log 100 is 2, because if you raise 10 to the power of 2 you get 100, and log 1000 is 3. Then the logs of any numbers between 100 and 1000 are somewhere between 2 and 3. Logs in base 10 basically count how many digits a number has in base 10. So as n gets bigger, $\log n$ always keeps getting bigger forever, but slowly. When n reaches a million, $\log n$ has still only made it up to 6.

This is part of what makes logarithms useful – they convert big numbers into small numbers so that we can manage them a bit better. There's a theory that when numbers get big beyond a certain point, our brains can't really process that sort of size anymore, so we switch to thinking logarithmically, observing how many digits the number has rather than how big it is. You might do this even without realizing this is called "logarithmically." It's why I told you above that the giant number 200! had 375 digits, because 375 is a number we can process; I was thinking logarithmically. By the time a number is that big, it hardly matters whether you've added 1 here or there, compared with the sheer size of the number. That's why in the rice example I became a bit cavalier and pretended there were 500 days in a year – I knew it was not significantly different from 365 in the grand scheme of things, and that all I really needed to do was pick a number with the same number of digits as 365. I was thinking logarithmically.

Logarithms grow more slowly than raising n to any fixed power.

Slow Growth

It's possible to grow so slowly that it looks like you're not growing at all, and yet you're actually still growing. Imagine being on a diet and deciding that you'll only have half a slice of cake. And then it's so delicious you decide to eat another third of a slice of cake. And then, because you're like me and you keep wanting more, but you're really trying not to get fat, you decide to eat more, but this time only a quarter of a slice. And then a fifth. And then a sixth. And so on forever. How much cake will you end up eating? After a while the new pieces you get will be essentially nonexistent, because after a million rounds of this you'll only be eating a millionth of a slice of cake, which is basically nothing, right?

Wrong. If you keep going like this forever, you will actually end up eating an *infinite amount of cake*. In fact, the amount you eat grows logarithmically. That is, slowly, and more and more slowly as it goes along, but still inexorably heading toward infinity. This is what a graph of log looks like:

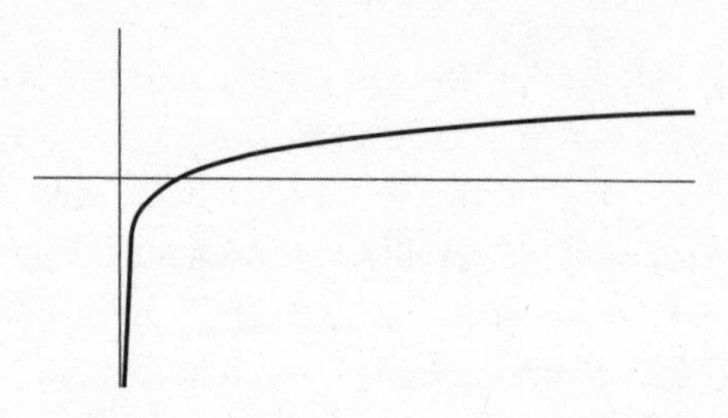

From the amount of graph I've drawn here you can't tell whether it's going to tail off completely flat or not, but it really doesn't. It's *unbounded*, which means that whatever number you think of, it will eventually get above that number. No matter how big you want it to be, it will always get bigger than that eventually. We will explain this in Chapter 16.

This example shows that we really have to be careful when thinking about infinite growth because our intuition can definitely lead us astray. How can we end up eating infinite cake if every slice is basically nothing? This is why some more rigorous mathematical precision is needed, to make sense of these confusing apparent contradictions, and to be able to tell, for sure, whether we're eating infinite cake or not.

12 Infinite Dimensions

Would you like to be able to travel in time? It sounds exciting, but it also sounds terrifying because of the dire consequences if you even slightly interfere with your own past. Those loops and potential paradoxes intrigue me and are the subject of some of my favorite fictional scenarios: the film *Back to the Future*, the book *The Time Traveler's Wife*, and the more recent film *Looper*, whose scenario was so mind warping to me that I had to read the synopsis on Wikipedia at the same time as watching the film.

Time travel has less potential for dangerous consequences if you just do it very briefly and don't interact with anyone while you're away. This might sound pointless, but it could be very useful for escaping if you're being chased by baddies. It consists of escaping *into a fourth dimension*. In this chapter we're going to think about how many dimensions there are in the world.

Imagine you're on a train and someone is trying to chase you. In order for them to catch your train, all they have to do is block it in front and behind. The train is stuck on a track so won't be able to go round the blockage – it can only move in one dimension, not two.

However, if you're in a car, you can drive round a blockage, so the baddies would have to block you in a whole circle around your car. At this point, to escape you need to be James Bond and press a button that turns your car into a plane so that you can fly out of the circular trap. You've escaped into the third dimension.

Now in order for them to catch you they'll need to cast an entire net around you in the sky. To escape this you'll need to use the fourth dimension . . .

We're so used to living in a three-dimensional world that it can be hard to imagine what more dimensions would be like. You might immediately think, "But there *aren't* more dimensions." This is true if we're thinking about physical dimensions, but that's only one way to think about dimensions, a particularly concrete and physical way. Concrete examples are useful for getting to grips with an idea but can have their limitations, just like understanding numbers via cookies is fine but that makes it hard to think about negative numbers, because "negative cookies" are hard to imagine.

One way to think about four dimensions is as a generalization of three dimensions. To understand how this generalization works, it's helpful to take a few more steps back so that we can get a bit of a running start. If we see how to go from one dimension to two, and then from two to three, it helps us to see how to go from three to four, and then from four to five, and then from any n to any $n + 1$, just like the child climbing up stairs or us climbing the ladder of infinities. Then maybe if we never stop, we'll get an infinite number of dimensions. (Sometimes I think this idea of generalizing forever is a form of mathematical optimism.)

If you're in a one-dimensional world, that means you're basically living on a straight line. Actually the line doesn't have to be physically straight, it just means there can be only one direction in which you can go (forward or backward counts as positive and negative of the same direction). Another way of thinking about it is that you can tell someone where you are by giving them just one coordinate. Whether a street is straight or curved, the houses can still be numbered in order. If you're on a circular path, you still only have to give one coordinate to someone for them to find you – you can just tell them how far round the circle you are. You *could* give them two coordinates, telling them how far north and how far east you are, but that would be inefficient and unnecessary. For example, instead of saying you're 50 meters north and 50 meters east, you could say you're 45 degrees around the circle, counterclockwise from some agreed starting point.

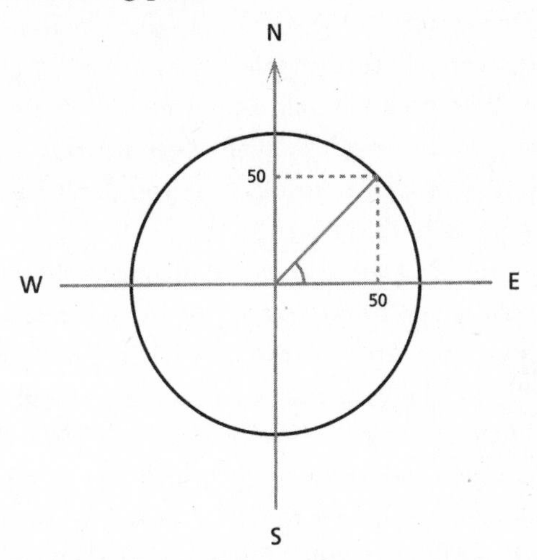

One-dimensional worlds are thus very straightforward. There's only one *variable*, which is how far along you are. Still, it's amazing how many times I've managed to get lost in a one-dimensional world. When I worked at the University of Nice, the department was circular and I could somehow never find my office. And I frequently get lost on a train when I'm trying to find my way back from the bathroom, because although abstractly forward and backward are part of the same dimension, in practice they're rather different. It doesn't help if I'm thinking about math at the same time as trying to find my seat.

A two-dimensional world enables you to escape from the one-dimensional world. This is what happens when you get off the train (or jump out of the window if you happen to be James Bond). The problem with the one-dimensional world is often all too apparent in Sheffield as the one-dimensional trams share the streets with the two-dimensional cars. (The trams and cars themselves are both three-dimensional of course; I'm referring to the number of dimensions in which they move.) If a car breaks down on the tram track, the other cars can just drive round it but the tram is completely stuck until the car is towed away. And it can't just go backward because then it would be going the wrong way up the street. To trap something in a one-dimensional world is very simple – you just need a blockage. That's why they used to build moats around castles, because then the way to the castle essentially becomes one-dimensional (the drawbridge) and easy to block. If you don't have a moat, you need a rampart all the way round.

You can tell if a world is two-dimensional because you'll need two coordinates to say where you are: you need to specify how far "along" you are, but also how far "across." This is why theater seats and plane seats have a row number and also a seat number. (In theory, because the number of seats is finite, you could just number them all starting from 1 and going up, but it would make it much harder to find your seat.)

The surface of the earth is a funny one because it *seems* to be three-dimensional but really it's only two-dimensional, because you only need two coordinates to specify where you are: longitude and latitude. This is not taking into account how far off the ground you might happen to be. The surface of a sphere is two-dimensional even though it only fits inside a three-dimensional universe. This is a hint that the notion of "dimension" is not as straightforward as it seems, even if we're talking about physical dimensions. This is just like the fact that a circle is one-dimensional, even though you need a two-dimensional piece of paper on which to draw it. A simple road going through the mountains is also one-dimensional, but it wiggles left and right as well as up and down so only fits inside three-dimensional space. The two dimensions of two-dimensional space are often cartesian, that is, X- and Y-coordinates like in the cartesian grid system of Chicago:

However, I was recently in Amsterdam and it occurred to me that the concentric canal system made the city into a polar coordinate grid, where instead of east/west and north/south, there's how far round you are (an angle) and how close to the center you are:

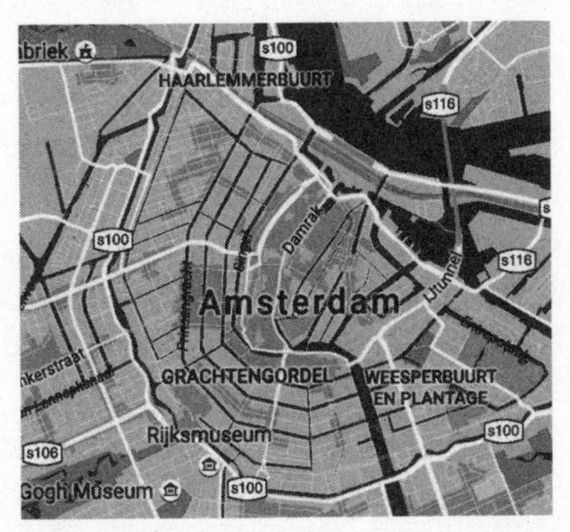

A three-dimensional world enables you to escape from the two-dimensional world. One of the reasons I love swimming underwater is that I get to experience three-dimensional freedom. I suppose this is one of the fun things about hang gliding or skydiving, except I'm much too scared to do that. Also I wouldn't feel quite so free about that third dimension as I'd be at the mercy of gravity. Three dimensions is why you only need a fence to keep in a cow, but you need an entire cage to keep in a bird. It's also why plane fights are more complicated than car chases. A plane needs three coordinates to specify its position: longitude, latitude, and also the height it's flying at. It's why you can catch a cow with a lasso (well, maybe *you* can't, but someone can), whereas for a fish or for Superman, you need a net (or some kryptonite).

A four-dimensional world enables you to escape from the three-dimensional world. If a fish had miraculous access to a fourth dimension, it could escape your net. If you were James Bond in a plane and surrounded by baddies, you could hit a button and escape using the fourth dimension.

Possible Fourth Dimensions

If you find this hard to understand, you can imagine time as a fourth dimension. This is a valid and in fact powerful way to think of time, put to great use in theoretical physics, but it is just one way of thinking of a fourth dimension, not the only way. What this means is that if you're captured, you could escape by time travel. This is a common theme in fiction involving time travel. In *Back to the Future* Marty escapes from the people who shot the Doc by time traveling (albeit accidentally). In *The Time Traveler's Wife* Henry is arrested but escapes by time traveling – also accidentally.

Another way of thinking of it is that time is another coordinate that you need to specify. If you're meeting up with someone, you need to specify where but also when. If you both go to the same coordinates but at different times, you won't meet up; you'll be at the same place but you won't be at the same place in "space-time."

Here's how you escape using time as the fourth dimension. First, let's think about how you escape from two dimensions using the third. If you've been fenced in, it's not hard to escape: you climb over the fence. The person who has fenced you in has mistakenly believed that you can only move in two dimensions, which is to say that you can't change your vertical coordinate. All you have to do is change your vertical coordinate briefly, get over the fence, and then your vertical coordinate can go back to being zero again now that your east and north coordinates are safely outside the fence.

If someone has caught you in a net, you'll need to change your time coordinate to get out. You can just briefly time-travel into yesterday, where you are not caught in a net and can move sideways a bit. Then you can time travel back into today, now that your physical coordinates are safely outside the net.

Another way to think of it is imagining that the fourth dimension is colors, so that you're only really in the same place as something if you have the same east, north, and vertical coordinates *and* you're the same color. You can change your color coordinate by painting yourself a different color. This means that for someone to trap you in a room with white walls, they'll have to paint you white. For you to escape you just have to get hold of some other color paint, say purple, paint yourself that color, and then you can walk straight through the white walls. You can then revert to whatever color you want when you're safely through the walls. It's a bit like an invisibility cloak, except those have only two states, visible and invisible, and even if you're invisible you can't necessarily walk through walls.

I recently explained all this to a musician, who immediately said, "Does that mean music is a fourth dimension, because it enables you to escape the three physical dimensions?" I would say yes, although it's a bit tricky because it doesn't exactly enable you to escape *permanently*. If you're trapped in a room, you can escape the sensation of being trapped by listening to music, but you can't use it to get yourself physically outside that room. If I'm on a train listening to my music and someone else is listening to some completely different music, it does seem like we're not in the same place at all, and if I've just been to a particularly powerful concert, I'll emerge into the street and feel like I'm not in the same dimension as all the people who were not there. It can also be like time travel as I can listen to a piece of music and immediately feel like I'm in another time (and place) that I particularly closely associate with that music.

You might wonder why it has to be about escaping. The reason is that if you're not escaping the previous dimensions, it's not really a new dimension. It's like the fact that you *could* use more coordinates to specify your position. You could say how far east you are, how far north you are, and also how far northeast you are. "I'm 4 miles east, 2 miles north, and $\sqrt{18}$ miles northeast."

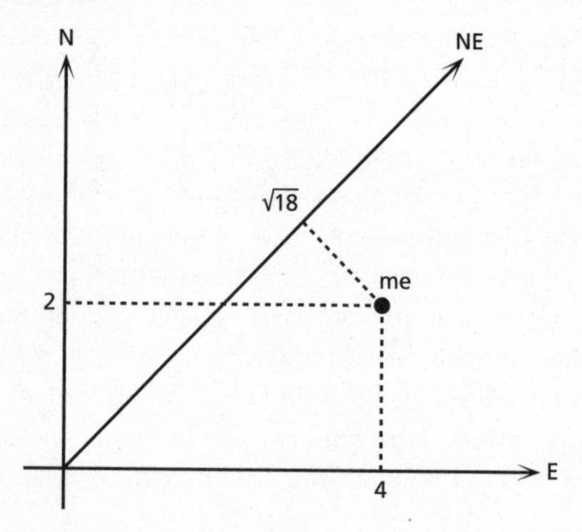

That would be accurate but redundant. In mathematics this is the notion of "independence" of the dimensions. If one of the dimensions can be expressed in terms of the others, it's not independent. This is linguistically evident in our example as "northeast" is a word made from "north" and "east."

Does Anything Really Have Four Dimensions?

As my research is in higher-dimensional category theory, I spend a lot of time talking to people about dimensions, both formally and informally. When I give nonspecialist talks, there will typically be someone in the audience who gets very upset and insists that it's all pointless because there simply *isn't* a fourth dimension, because we simply do live in a three-dimensional world.

That is correct physically. But once you start thinking of dimensions in terms of independent coordinates, things completely change. I said earlier that once we can imagine something in mathematics, it exists. As long as we can imagine more than three independent ideas, then we have created an idea space with more than three dimensions. And I am sure you have had more than three independent ideas in your life. Just like Molière's *bourgeois gentilhomme* who discovered he had been speaking in prose all his life without realizing it, you have probably been thinking in many more than three dimensions for much of your life, possibly without realizing it.

Coordinates are, after all, just a string of numbers. The numbers can represent anything that is on some kind of measurable scale. Distance is one such thing that is on a measurable scale. To take those words rather literally, weight (or rather mass) is something else you can measure on a scale. The last time I went to the doctor, they recorded my age, height, weight, pulse, and blood pressure. Given that blood pressure is two numbers, that's six numbers right there. If you wanted to plot a graph of everyone's information, you'd need a six-dimensional graph.

I've just spent several days practicing making French macarons. It was very confusing because there were so many variables. Although there aren't many ingredients, you still have to decide how much icing sugar, caster sugar, and ground almonds to use per 100 g of egg whites. Then you have to decide how long to whisk the egg whites, how long to fold the batter, how large to pipe the circles, how long to let them stand, what temperature to have the oven, and how long to bake them. That's a nine-dimensional space. And that's not even taking into account the state of confusion I get into when I'm trying to buy eggs, which have their own variables: size, color, free range or not, organic or not, vegetarian or not. I want them to be large, brown, free range, organic, and vegetarian, but I frequently get to the checkout and discover I forgot to check one of those variables.

Every time you compare things according to a list of criteria, you're effectively looking at a space with that many dimensions. From this point of view, it would be quite rare to have as few as three dimensions. Once you've bothered going as far as listing criteria, there are probably going to be more than three. This might be true even if you don't notice yourself listing them: every time I buy a plane ticket I weigh up the price, the airline, the schedule, the number of stops, and the airports, but I do it in my head.

Robotic Arms

You might think this is all very well, but is there ever a need to *study* those higher-dimensional spaces? Do we even need to know that they're there? You've been getting along with your life perfectly well without knowing you're thinking about higher-dimensional spaces; you also probably spoke for many years without knowing you were speaking in prose.

One situation in which higher-dimensional spaces are studied is space-time, which consists of our normal three-dimensional space together with time as a fourth dimension. Einstein's theory of relativity involves this four-dimensional space being curved.

Something a little more concrete than that is the study of robotic arms. Robotic arms are used all over the place, including factories, outer space, keyhole surgery, and arcade games. The arm itself is moving inside three-dimensional space, but to study its range of movement you have to consider what each hinge or socket is doing at any given moment. Each hinge is one variable, so you end up with a space with as many dimensions as there are hinges, or more if they are more complex sockets.

Look at your own arm for a moment. If you sit still and wave your hand around, your hand appears to be moving in three-dimensional space. How is it doing that? There are various hinges and joints giving different specific types of motion.

- If you wave your arm around but keep everything rigid from your upper arm to your hand, you'll see that your shoulder joint all by itself gives you two dimensions of motion: to specify where your upper arm is in relation to your body, you'll need two coordinates, up-down and front-back.
- Your elbow gives you one coordinate, for the angle your forearm is making with your upper arm.
- Your wrist gives you two coordinates for the position your hand is in relative to your forearm: one coordinate for up-down and another for left-right.
- There's also forearm rotation, which you might also think of as hand position: the hand can rotate from palm up to palm down.
- There's also the rotation of your upper arm: you can keep your upper arm in the same place relative to your body and fix the angle your forearm makes with your upper arm, but still rotate. This is more or less what happens if you're waving at someone, the type of wave where your hand makes an arc in the air, not the type where you flap your hand.

That's seven dimensions. Is my hand really moving in seven-dimensional space?

If you've ever worn a dress, you may have spent some time trying to do the zipper up the back. If you've never worn a dress, perhaps you've met a similar issue trying to put sunscreen on your own back. What happens with the zipper, typically, is that you can do up the bottom portion with your hands behind your back, but at a certain point you get stuck and have to switch configuration so that your hand is going over your shoulder and can then pull the zipper up to the top. If you're not flexible enough, you can't reach and need help. It's similar if you're putting sunscreen on your back. You can do your lower back with your hands behind your back, but you only get up to a certain point, perhaps around your shoulder blades. Then you have to switch to your hands being over your shoulders to do your shoulders and upper back. If you're not flexible enough, there's a gap between the lower part and upper part where you'll get sunburned.

What's happened here is that there's a boundary of where your hand can reach in the lower position. If you're flexible enough, that will be right next to where your hand can get from the upper position, maybe even with a little overlap. In both cases your hand is reaching the same place inside three-dimensional space. However, your arm is in a completely different configuration and has to travel a long way to get between those two configurations. This means that those two places are very far apart in the full seven-dimensional space of arm configurations. This is the kind of thing we need to know when we're designing robotic arms. We do need to know where the "hand" can go in ordinary three-dimensional space, but it's important to know which positions are close together in the full higher-dimensional space of motion of the arm. If you need to gently move something a tiny distance during keyhole surgery, it will be no good if the robotic arm has to do a full reconfiguration in order to get there.

Reducing Dimensions

Thinking about higher-dimensional space is difficult. That's why there are whole branches of math devoted to it. That's why in life and in math we often try to reduce the number of dimensions we're thinking about, to make life easier. There are various ways of doing this.

One way is just to ignore some dimensions. For example, if you're evaluating something according to a list of criteria, you might realize you have far too many criteria and you should focus on the important ones. Ignoring some of the least important ones instantly reduces the number of dimensions you're thinking about and makes the decision easier.

You can also ignore some dimensions just temporarily, by fixing some of the variables. When I was trying to perfect my macarons, I eventually studied them one dimension at a time. I made one batch and baked them at all sorts of different temperatures. Then I fixed the temperature and made one batch where I piped the mixture at progressive stages of mixing. Then I fixed both of those things and changed the amount of ground almonds a little bit. This is like taking normal two-dimensional space and fixing the y-coordinate to be 2, say. Then the only thing you'll have left is this straight line, where x can be anything but y has to be 2:

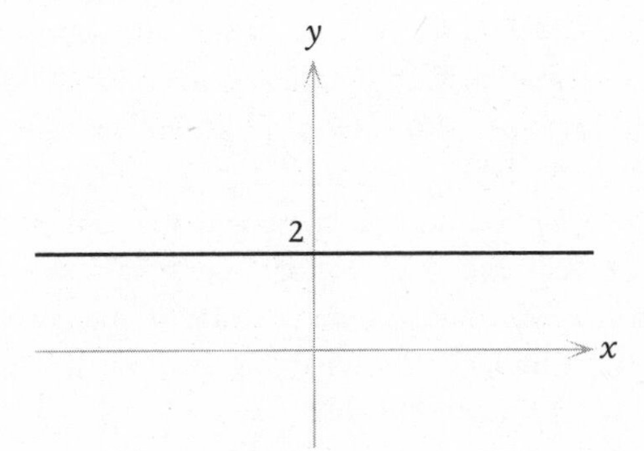

We have turned the two-dimensional plane into a one-dimensional line. This is not that different from entirely forgetting a variable: if you entirely forget a variable, it's like reducing yourself to the x-axis, which is like fixing y to be 0.

In the case of my macarons, it was complicated by the fact that I didn't know if the best temperature for one recipe would be the best temperature for another recipe. So then after varying the amount of ground almonds, I had to try all the different temperatures again. As you can imagine, this investigation went on rather a long time.

A slightly more subtle (and I would say sneaky) way we often reduce the number of dimensions is by conflating two dimensions into one rather than just forgetting one. This happens all the time in questionnaires when they ask you questions like "I prefer socializing in large groups rather than one-to-one" and then you have to say if you agree or disagree on a scale like this:

o	o	o	o	o
strongly disagree	disagree	neither	agree	strongly agree

This is supposedly more nuanced than just saying yes or no, because instead of things being black or white, we've allowed for gray areas in between. However, these sorts of questions frustrate me because I often find myself strongly agreeing *and* strongly disagreeing in equal measure. It seems that I'm supposed to select "neither," but that doesn't feel right because "neither" sounds like I don't really care either way, whereas actually on some days I *really* want to be in a big group and on other days I *really* want to be in a one-to-one situation. This is because it's really a two-dimensional question: How much do you like socializing in large groups, and how much do you like socializing one-to-one? This gives us two variables, and we could plot it on a graph like this:

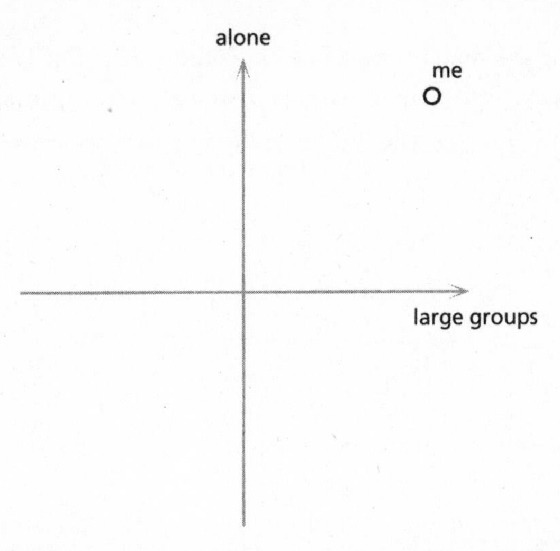

I would be somewhere up in the far top right as I like doing both. Someone who *hates* doing both equally would be down in the bottom left. In fact the whole diagonal is different ways of liking or disliking both equally.

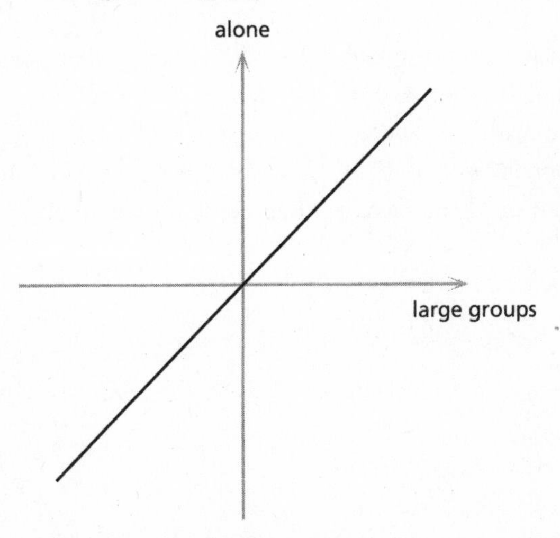

Mathematically this is the line where $x = y$. In the original scale, this line has been squashed to a single point "neither." What the questionnaire has done is assume that socializing in large groups and small groups are "opposites" of each other, and that the amount you like one automatically tells us how much you dislike the other. It's like a zero-sum game. Mathematically the zero-sum game means $x + y = 0$, which is this line:

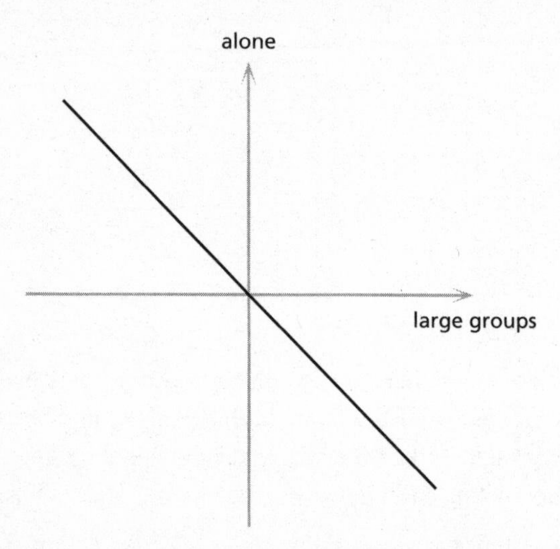

Our situation is more like a "four-sum game" where the total is always 4. To get 4, I'm considering that you can agree or disagree with the statement on a scale of 0 to 4. Then the questionnaire seems to be assuming that however much you agree out of 4, you have to disagree by the difference, like this:

alone	4	3	2	1	0
large groups	0	1	2	3	4
	o	o	o	o	o
	strongly disagree	disagree	neither	agree	strongly agree

If we plot these points on the graph, it gives this:

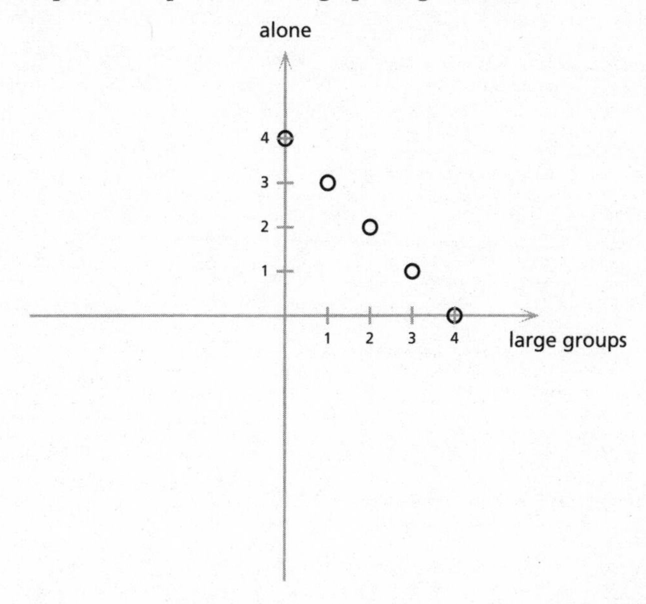

Sometimes the questionnaire tries to be more nuanced and admits that these things are on a continuum, and they allow you to mark any place on a line from "agree" to "disagree":

In that case we're taking the entire line $x + y = 4$, which is here:

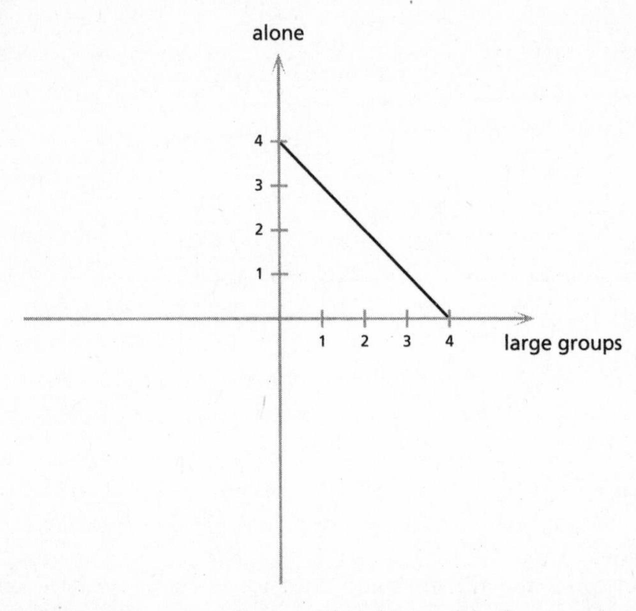

You can pick any point on this line, and if you take its x-coordinate and y-coordinate, you'll find that they add up to 4. So we've reduced two dimensions to a one-dimensional line in a slightly odd way.

Another way of considering the sliding scale above is that we're really considering the *ratio* of how much we like one type of socializing compared with the other. In that case we're taking the x and y coordinates and reducing them to $\frac{x}{y}$. This is an even more complicated way of reducing two dimensions to one. There are online tools where you can get a rendering of a three-dimensional graph. You can try doing $z = \frac{x}{y}$ to try and show you what value you get for each position in the two-dimensional plane. It is very hard to visualize.

This reduction of dimensions also happens with political beliefs, where we tend to talk about how "left-wing" or "right-wing" people are, which is an extreme oversimplification. The website www.politicalcompass.org points out that there are really two distinct variables, economic views and social views, making a two-dimensional political graph, which we tend to squash into one dimension. There are always advantages and disadvantages to simplification. The advantage is that simpler things are easier to grasp, but the disadvantage is that you lose information. Those things should always be weighed up, and we should at least be aware that we're doing it.

Really, political beliefs sit in a much higher-dimensional space than even two dimensions, because there's no limit to the number of variables we can consider when characterizing our political beliefs. Perhaps we've even gotten ourselves into an infinite-dimensional space, but we have to reduce it so that we can attempt to organize ourselves into groups of people who broadly agree with one another. The trouble is that we tend to do it by conflating dimensions rather than by ignoring them. Instead of deciding that certain criteria are unimportant, we pretend that they're *linked*. Mathematically those are very different ways of reducing dimensions.

A Continuum of Dimensions

Sometimes when I'm doing a "pros and cons" type evaluation of a situation, I discover that the criteria are too difficult to separate out. For example, you might be thinking about short-term benefits and long-term benefits. Then you start wondering about medium-term benefits. I've also started considering micro-term benefits (instant gratification). Then where does short term turn into medium term? Where are the boundaries between those things?

Maybe you're evaluating different jobs, and you consider job satisfaction and also self-esteem. But those two overlap because self-esteem in your job gives you job satisfaction. Maybe self-esteem is just one aspect of job satisfaction? You start splitting job satisfaction up into more and more specific criteria, but as you do so the boundaries between them get more and more blurry.

Eventually you realize that your criteria themselves are on a continuum as well. What this means is that not only are you evaluating each individual criterion on a continuum, but there's an entire continuum's worth of criteria. You're in a space with uncountably many dimensions. No wonder decisions are difficult.

13

Infinite-Dimensional Categories

A friend of mine posted something on social media about a delicious croissant sandwich, and I immediately imagined a croissant inside a sandwich. I soon realized she meant that the croissant was the outside part of the sandwich, not the filling.

Have you ever thought about making a "sandwich sandwich"? This is a sandwich in which the filling is itself a sandwich. Perhaps there should be lettuce between the extra slices of bread? Some people insist that lettuce plays a vital – indeed structural – role in a sandwich, protecting the bread from being made soggy by the filling (although personally I would never eat lettuce by choice). In that case an x sandwich would consist of

bread
lettuce
x (filling)
lettuce
bread

For example, a chicken sandwich would be

bread
lettuce
chicken
lettuce
bread

in which case a chicken sandwich sandwich would be

bread
lettuce
chicken sandwich
lettuce
bread

which, if we write it all out, becomes

bread
lettuce
bread
lettuce
chicken
lettuce
bread
lettuce
bread

This is a bit like my favorite cake, the "iterated Battenberg cake," where you make a Battenberg cake

but this time so that each of the individual cakes is itself a Battenberg cake.

Or perhaps there should also be marzipan around the small Battenberg cakes inside the big one?

Both of these are examples of making something out of, essentially, itself. Mathematics is particularly good at making things out of itself, like how higher-dimensional spaces are built up from lower-dimensional spaces. This is because mathematics deals with abstract ideas like space and dimensions and infinity, and is itself an abstract idea. Physical objects don't behave like that. If you stick a lot of birds together, you won't get a new bird; that just isn't how birds work. So when is something abstract enough to be iterated?

Lego Lego

Imagine making giant Lego blocks out of Lego. This works because there's something a bit abstract about Lego. You can't make a giant bird out of birds – not a real one anyway. But you could make a giant bird model out of smaller bird models, because a model of a bird is abstract enough, whereas a real bird is not. There are artists who build portraits of people in a mosaic-like fashion, using individual portraits as tiles. You can make a portrait out of portraits, because a portrait is also abstract enough, whereas the actual human being is not.

When I was little I learned to program on a Spectrum, just like a whole generation of British programmers, mathematicians, and others in that-way-inclined families. I have already talked about my favorite "infinite HELLO" program. I never tire of describing the concept of a Spectrum to those who haven't met one, because it's so simple and so extraordinary. I'll describe it now just in case: it was one of the earliest (or possibly the earliest) version of a home computer. It consisted of basically just a keyboard – a tiny little thing with rubbery keys. It had no screen – you plugged it into your television set. To save things, you used ordinary cassette tapes and an ordinary cassette recorder – you just had to press record on the machine and hit "save" on the Spectrum and it would record your program as funny little sounds. I do a very plausible impression of these sounds by singing them – plausible enough, at least, that anyone who grew up with a Spectrum will be hit by a wave of nostalgia and possibly a fit of giggles. To load the program back again, you pressed play on the tape machine, and hit "load" on the Spectrum and it would play back those funny sounds and miraculously load the program. There were also some quite exciting games that you could buy on cassette tapes too (and some I found less exciting, like the flight simulator that just didn't seem to do anything, perhaps because I was no good at it).

The printer was an adorable little thing that plugged in, I forget how. The really memorable part was that it used rolls of silver paper, like toilet rolls, but smaller. It was silver on the printing side, and I think it printed using heat. There was only one font in only one size, and it was all equally spaced, like Courier.

I liked making signs saying things like EUGENIA'S ROOM, but there was no way of changing the font size, so to make it big enough the most fun thing to do was to make each letter out of letters, as in:

EEEEEEEEEEE
EEEEEEEEEEE
EEEE
EEEEEEEE
EEEEEEEE
EEEE
EEEEEEEEEEE
EEEEEEEEEEE

and so on. This works with letters, because they, too, are abstract things. Everything in mathematics is abstract, so you always build more mathematical things out of other mathematical things, even if they're not exactly the same thing. For example, you could make an *E* out of *A*'s, which is oddly counterintuitive.

AAAAAAAAAA
AAAAAAAAAA
AAA
AAAAAA
AAAAAA
AAA
AAAAAAAAAA
AAAAAAAAAA

You can do this sort of building up with mathematical objects, as they are abstract. For example, you might remember what matrices are. They look like this:

$$\begin{pmatrix} 1 & 0 \\ 3 & 2 \end{pmatrix}$$

This is a matrix built from numbers, but we could build matrices from other things as well. When I took math A-level, graphical calculators had just arrived and, miracle of miracles, they could also do matrices. The people writing A-level exams quickly caught on to this though, and made us do matrices with letters instead of numbers, to foil our attempts to use our calculators. A matrix made of letters looks like this:

$$\begin{pmatrix} a & b \\ c & d \end{pmatrix}$$

But we could just as well make a matrix out of birds.

Or a matrix out of . . . matrices.

$$\begin{pmatrix} \begin{pmatrix} 1 & 0 \\ 3 & 2 \end{pmatrix} & \begin{pmatrix} 2 & 5 \\ 1 & 1 \end{pmatrix} \\ \begin{pmatrix} 3 & 2 \\ 0 & 1 \end{pmatrix} & \begin{pmatrix} 1 & 2 \\ 4 & 3 \end{pmatrix} \end{pmatrix}$$

Matrices are abstract enough that we can make matrices of matrices. Mathematics is abstract enough that we can always make more mathematics out of mathematics. Christopher Danielson's marvelous book *Which One Doesn't Belong?* includes one "Which One Doesn't Belong?" example made up of Which One Doesn't Belong? examples: which "Which One Doesn't Belong?" doesn't belong?

In my previous book I talked about my field of research, category theory, as "the mathematics of mathematics." What about the mathematics of category theory? It's still category theory, but it gets more and more dimensions. Here's how. It stems from the fact that category theory is all about relationships between things. And what if those "things" are themselves relationships?

Planes, Trains, and Automobiles

At the very beginning of this book I compared journeys by plane with boat trips. If I'm flying across the Atlantic, I'm not doing it for the view – there's not much to see for most of the time. Maybe one day I'll take a boat across the Atlantic, but I usually need to get there more quickly than that.

When you're going on vacation, do you consider different ways of getting there, or do you always go by plane? Perhaps you're determined to go somewhere sunny, in which case, if you live in the UK, you're very likely to need to fly.* Or sit on a train for a very long time.

When I was little we used to take the car to France, visit vineyards, buy wine, and drive back again. Living near Brighton, it was quite convenient to drive to a ferry crossing point. There were two main options: Newhaven–Dieppe or Dover–Calais. It was a much shorter drive to Newhaven, but a much longer crossing. I don't know which was more expensive; I was too young at the time to ask such questions.

At some point when I was still quite young, the hovercraft became a viable option. This apparently miraculous device flew across the water. Imagine my disappointment the first time I went on one, when I found it didn't feel like flying at all. It felt exactly like being on a boat, except if I remember correctly, it was much more susceptible to rough seas and we were all horribly seasick.

Taking the car is a practical choice when you have small children – apart from being cheaper, you can also throw all your paraphernalia in the car and still have plenty of space for shopping, especially if your car has a large trunk like our old Saab did. Once you've decided to drive somewhere, you still might compare different routes. If you get directions from Google, it will offer you that. You can compare the time, the distance, the sorts of roads you'll be driving on. If you're me, you take the simplest route, on major roads, to avoid getting lost, but many other people would pick the shortest or the quickest.

All of this is about comparing paths from one place to another. A route from *A* to *B* is a form of relationship between those two places, and there can be many different routes between the same two places. Now if we're looking at relationships between different routes, that is relationships between relationships.

Category theory is all about relationships between things, and now we have a notion of *dimension* in category theory.

* If you ignore relationships between things and just consider everything as if in a vacuum, that's zero-dimensional.
* If you allow relationships between things, or paths between places, that's one-dimensional.
* If you take into account relationships between those relationships, that's two-dimensional.
* If you have relationships between the relationships between the relationships, that's three-dimensional.

. . . And so on.

The idea is that if we've decided we should take relationships between things into account, why not relationships between those relationships as well? Studying relationships between things is about putting things in context. If we want to study things in context, shouldn't we also study relationships in context? We might compare different ways of going on vacation and find that flying is quicker but driving is more expensive. Next we have to compare those relationships: Is speed more important than cost? Or perhaps you're gossiping about other people's relationships (surely not). You observe that one couple has a lot of fun together but also argues a lot, whereas another couple never argues but they don't seem to have fun either. Then you can compare whether having fun is more important than not arguing.

Category theory studies relationships between things and builds on this in various ways: characterizing things by what properties they have, finding the pond in which things are the biggest fish, putting things in context, expressing subtle notions of things being "more or less the same." The higher-dimensional version does this whole thing to the relationships themselves. It's another level of abstraction, and it takes us into *higher-dimensional category theory*.

In category theory we draw relationships between things as arrows, like this:

$$A \xrightarrow{F} B$$

Perhaps A is where we live and B is where we're going on holiday, and F is how we're going to get there. Our journey is a "relationship" between our home and our vacation destination. Now maybe we have two possible ways of getting there, F and G, and a way of comparing them. In higher-dimensional category theory we would draw it as arrows like this:

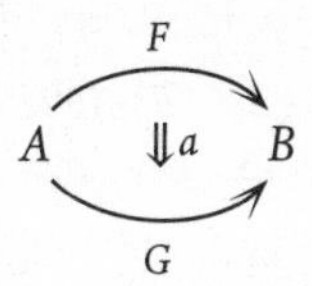

Here the double arrow labeled a is the relationship between route F and route G, which could encode a comparison of cost, or time, or fun, or the views along the way.

We can then build these up into big diagrams like the one below, which is a genuine diagram taken from my research.

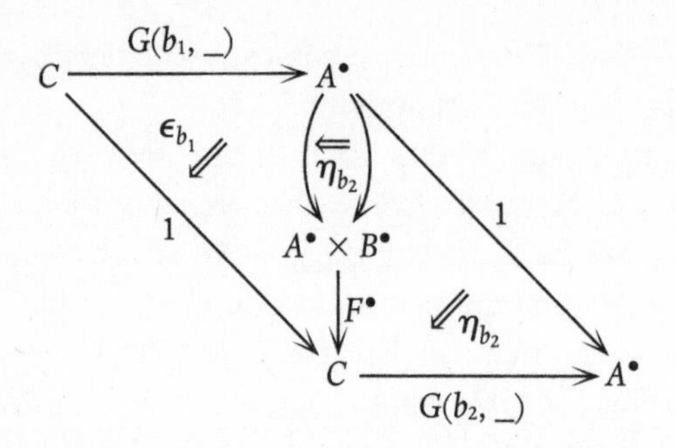

Higher-dimensional category theory studies these higher-dimensional structures that just never stop arising until you say stop, enough, this is enough dimensions. But if you don't say that, they will keep arising forever and we have *infinite-dimensional categories*. We're going to see that in some ways these turn out to be less complicated than the finite-dimensional ones, because they're more "naturally" arising. In mathematics, especially category theory, things that grow up organically without us imposing restrictions are often less complicated. It's the artificial restrictions we impose that make things more practical but also more complicated. In a way this is like the fact that the adult world has many more restrictions than a child's world of imagination and idealism. The restrictions make it more practical but also more complicated to navigate. Idealism is easy when you're not the one having to turn it into some workable logistics.

In category theory there is always a tension between the idealism and the logistics. There are many structures that naturally want to have infinite dimensions, but that is too impractical, so we try and think about them in the context of just a finite number of dimensions and struggle with the consequences of making these logistics workable. Just like in the previous chapter where we saw different ways of reducing dimensions when thinking about criteria for evaluating a situation, category theory has different ways of reducing the number of dimensions. Each way causes different problems.

You might wonder why we can't just forget some like in the last chapter. We'll come back to this later, after thinking about how mathematics propagates. Just as every dimension of relationship we consider creates a new dimension of relationships to consider, every question we answer creates more questions.

Climb Every Mountain

The film *Alive* tells the horrific story of the famous 1972 plane crash of Uruguayan Air Force flight 571 in the Andes. The plane crashed into a mountainside in bad weather. In the film, we see survivors picking up a radio signal and hearing a news report that the search for survivors has been called off. They determine that a small group of the fittest ones will walk out of the mountains to get help; a whole Uruguayan rugby team was on the plane, so some of the survivors were pretty fit.

We see three of them set off with what they think are enough supplies to get them to safety. They must first cross a mountain ridge that they can see in the distance before descending to the valley. But when they get to the top of that ridge, all they can see are higher mountains in front of them. The higher mountains were previously blocked out by the ridge, but now they have climbed high enough to see what it is they really have to cross.

At this point they realize they don't have enough supplies for three of them, so one of them hands over his supplies and heads back to the crash site to wait for rescue. The scenario is unthinkable, but they do make it into the valley and the remaining survivors are rescued.

Mathematics involves nothing so horrific, but it is like an unending series of mountain ranges. Every time you conquer one, you have a moment of elation and then you realize that conquering this peak has just enabled you to see the higher peaks ahead, a bit like when I reached the top of my climb in

New Mexico and only then saw the extended ridge ahead of me. Or when I was swimming along the curved beach and each bit of curve just revealed more beach.

This is inevitable. In fact, it's part of the power of mathematics. It's because the concepts we study can always be built into bigger ones. Those, in turn, can be built into bigger ones, and so on forever. This is because of the abstract nature of it. If we are climbing mountains, we don't build more mountains out of the ones we've already climbed – we just see more that are there. In math, we keep building bigger and bigger ones.

It's not that we do it willfully; it's that the methods we develop for studying things in math are actually new pieces of math. To study those, we create new pieces of math that then need to be studied. This doesn't happen if you're studying birds: the methods you develop for studying birds aren't themselves birds.

This is how category theory arose, as a new piece of math to study math. In a way, category theory is an ultimate abstraction. To study the world abstractly you use science; to study science abstractly you use math; to study math abstractly you use category theory. Each step is a further level of abstraction. But to study category theory abstractly you use category theory.

Why Is It Hard?

You might be wondering if we can just add dimensions to our theory and be done with it. What might that mean? To specify a point in two-dimensional space we just have to give two coordinates: an x-coordinate and a y-coordinate. To specify a point in three-dimensional space we just add another coordinate. We can keep adding as many coordinates as we want to get as many dimensions as we want, even if we don't know what those dimensions look like *in space*.

Coordinates, and dimensions, don't have to refer to physical space: any time we have four independent variables, we have four dimensions. They don't have to be in space or time, as we've seen in the previous chapter with the idea of dimensions arising from *criteria* we're using to evaluate something, or from the hinges of a robotic arm.

Computers crashing can be studied in this way. An abstract (as opposed to physical) space can be constructed consisting of various configurations that the computer can get into, according to different variables, and a path leading up to a dead end will result in a crash. The space can then be studied using *topology*, the branch of math that studies the general shape of space. In topology, a ring doughnut has the same general shape as a coffee cup (with a handle) because they both have one hole, which is basically all that counts.

So why is all of this hard? Computers do, alas, still crash, despite our best efforts. The reason is that unfortunately it's all a bit more subtle than I've described. We don't just add dimensions to our categories by adding relationships between relationships and so on; we have to add *axioms*, or laws governing these relationships. And the more dimensions we add, the harder it is to know what these axioms should be.

One problem that comes up is the question of "associativity." Remember, in the most common case this means things like

$$(3 + 5) + 5 = 3 + (5 + 5)$$

– it doesn't matter where we put the brackets in a calculation. This makes it all very convenient, but in situations with objects more complicated than numbers, it's not genuinely true anymore. I've previously written about examples involving eggs, sugar, and milk where it's not true.

$$(\text{egg yolks} + \text{sugar}) + \text{milk}$$

makes custard but

egg yolks + (sugar + milk)

does not.

But here's a mathematical example. The field of topology studies the shape of space, and the field of *algebraic* topology studies it by looking at journeys through the space. This modern way of doing this is using a category. A category is a mathematical structure with

* objects
* some relationships between the objects, expressed as arrows between them
* a way of combining relationships if the arrows point in the same direction, so if you have arrows like this:

$$A \xrightarrow{f} B \xrightarrow{g} C$$

you get to combine them into one like this:

$$A \xrightarrow{g \circ f} C.$$

Now in this case our objects are all the possible places in the space, and our arrows are journeys from one place to another. We have a way of combining them because if you have a potential journey from A to B and a journey from B to C, then you can do one followed by the other and get a longer journey from A to C.

Now usually when we think about getting from A to B, we also think about how long it's going to take, in minutes or hours or something. In abstract mathematical space we don't really have units of distance, and we don't really have units of time either. They're both just numbers. If you're like me, the kind of person who lost marks by not bothering about units in tests at school, you'll be relieved by this absence of units.

To make it easier to compare journeys, all journeys are considered to take time 1. This is like when we do percentages and we think of everything as being out of 100 so that we can compare things. We're just bothered about what route we take and how long we spend on each part of the journey relative to the whole. So we "standardize" the overall time it takes to 1.

We know that if we have a journey from A to B and another from B to C we can stick them together to get from A to C. The trouble is that if we just stick them together, the total length of time will be 2 instead of 1. So in order to "standardize" this journey to get an arrow $A \longrightarrow C$, we have to pretend we're doing each journey twice as fast. So we spend half the time going from A to B and half the time going from B to C.

This is fine, but the trouble comes when we try to do three journeys:

$$A \xrightarrow{f} B \xrightarrow{g} C \xrightarrow{h} D$$

You might think it would be sensible to do each journey three times as fast, and it would. But this is not how associativity works. We can't just invent a new way of sticking three things together – we have to use the original way of sticking two things together, and do it with the brackets in different places, to see what happens. That is, if we make the composite journey

$$A \xrightarrow{f} B \xrightarrow{g} C$$

first and then stick h on the end, it should be the same as making

$$B \xrightarrow{g} C \xrightarrow{h} D$$

first, and sticking f on the beginning. The trouble is, *this isn't quite true*. Because the first way round, when we stick f and g together, we do each twice as fast, so we end up spending the following amounts of time doing them:

journey	time spent
f	$\frac{1}{2}$
g	$\frac{1}{2}$

Now, when we stick h on the end, we have to do *everything* twice as fast, so we have to spend half the time doing f and g in total, and half the time doing h. This means we spend the following amounts of time:

journey	time spent
f	$\frac{1}{4}$
g	$\frac{1}{4}$
h	$\frac{1}{2}$

But if we start by sticking the journeys g and h together, we'll do each of those twice as fast:

journey	time spent
g	$\frac{1}{2}$
h	$\frac{1}{2}$

and then when we stick f on the beginning, we'll spend half the time doing f and half the time doing g and h put together:

journey	time spent
f	$\frac{1}{2}$
g	$\frac{1}{4}$
h	$\frac{1}{4}$

If we compare these two tables, we see that we do not get the same answer. In the first way round, we spent only a quarter of the time on f and g and half the time on h, but the second way round, we spent half the time on f and only a quarter on g and *h*. So this method of composition is *not associative*. It really matters where we put the brackets.

There are various ways that mathematics deals with this, depending on the balance between practical logistics and ideals of abstraction. The situation can be approximated to get something more practical, but then some subtlety is lost. One way of doing it is to impose a stronger notion of sameness on the journeys, so that those two versions where you spend different amounts of time on them count as the same. This leads to a field of research called *homotopy theory*, because the notion of sameness involved is called homotopy.

Another way of dealing with it is to keep track of those slight differences using trees similar to the ones I've drawn elsewhere, to take into account the different possible positions of the brackets. So these two trees depict some different positions:

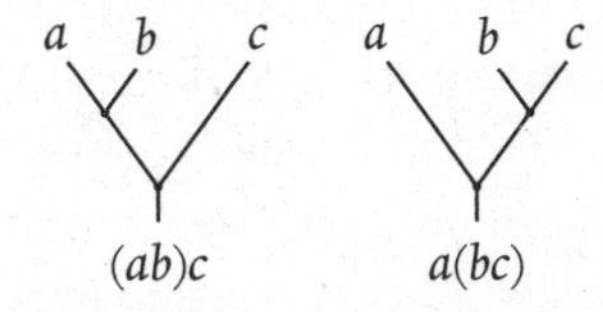

This leads to a field of research called *operad theory*, because the trees form an algebraic structure called an operad.

Another way of dealing with it is to look at the relationship between those journeys. This leads to higher-dimensional category theory. And then something curious happens with the higher dimensions. We can keep looking at the relationship between these journeys (or "paths" as they're formally called in mathematics), and the relationships between those, and between those, and so on. At any point we can decide we're tired and want to stop thinking about more dimensions, at which point we impose a notion of sameness at that top dimension so that we don't have to think about any more relations. This is a bit like deciding that certain criteria are not very important so you're going to ignore them and consider two situations to be the same, even if they differ according to those criteria. For example, you might decide that actually when you're buying eggs you don't mind if they're brown or white.

The trouble with doing this in category theory is that this has consequences for the axioms that we have to think about. It turns out that in some ways it is easier *never to stop* thinking about more dimensions, because we never have to deal with the consequences of forcing things to count as "the same" when they're not *really* the same.

This might sound counterintuitive, but think of it like this. Imagine if we decided there should be a biggest number: that's it, big numbers are too complicated, and there are going to be no numbers bigger than 1000. It would cause dire consequences because it would be so unnatural. We really *need* an infinite supply of bigger and bigger numbers, even if we don't ever use them.

With the dimensions in category theory, I like to think of it like this. With every dimension we add another layer of subtlety so that we can postpone the issue of forcing things to be the same and deciding what axioms to impose. If we have an infinite number of dimensions, we can postpone this forever. If we were immortal, we could procrastinate *forever*.

14 The Infinitesimally Small

I still get excited flying into a city and looking at the buildings from the sky. It hasn't stopped amazing me that the buildings look so big close up but so tiny from up in the plane. Manhattan is particularly surreal to me, as the skyscrapers look so squashed and crammed together on the little island. The old flight path into Kai Tak airport in Hong Kong was terrifyingly dramatic, as the planes had to fly so close to the buildings. One minute the skyscrapers would be tiny sprouting pins, and the next minute you'd be looking into someone's flat seeing them do the laundry.

Most of what we have looked at so far involves things growing, or infinite things appearing because each one gives rise to another one. However, we can quickly flip this scale and think about things being very tiny instead. This gives a whole different way of getting infinitely many things, by thinking about infinitely many infinitesimally small things. Does this feel like cheating? Sometimes mathematical advances happen by just looking at something in a slightly different way, which doesn't mean building something new or going somewhere different, it just means changing your perspective and opening up huge new possibilities as a result. This particular insight leads to calculus and hence the understanding of anything curved, anything in motion, anything fluid or continuously changing.

Few things in our world do not fit these descriptions. Computers are largely digital, meaning that they are divided into clearly defined bits that change in clearly defined steps rather than continuously, but they still involve electricity, which does fall into the "continuously changing" realm. One of the few things I can see in front of me right now that doesn't obviously involve calculus is the table I'm working on. It's true that tables existed long before calculus, but still, this particular table was made in an Ikea factory, which definitely involves calculus. My point is that although the study of infinity might seem abstract and out of this world both literally and figuratively ("figerally," as one of my friends wryly puts it), it led to the field of calculus, which is inextricably part of almost every aspect of modern life.

The starting point for all this is thinking about things that are "infinitely close together." When we draw a circle digitally on a computer, or type the letter o, it looks smooth and joined up, but if we zoom in enough it will eventually become pixelated. Here's a close-up of a letter o on my computer screen.

We see that it's just a finite number of square dots masquerading as a joined-up curve. In this picture my computer has faked the curve a bit better by including some shades of gray. The computer screen has to do this, as it can only understand individual dots, it only has a finite number of them to work with, and they are all a fixed, measurable size.

But what about our brains? The idea of calculus is that our brains can do better than that, in principle, because we can deal with infinitely many things, and we can deal with them being infinitely small. This is what we're going to investigate now.

Once I was helping with math at Park Street Primary School

in Cambridge. I was helping two six-year-olds understand symmetry. I got them to draw the lines of symmetry on some triangles, and then on a square, and then a pentagon, and then a hexagon. One of my favorite moments was when one of them said, "I know an octagon has eight sides because OCT stands for OCTOPUS." Anyway, I eventually gave them a circle. One of them drew this line

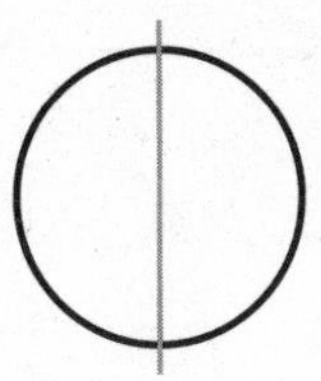

and then the other drew this line

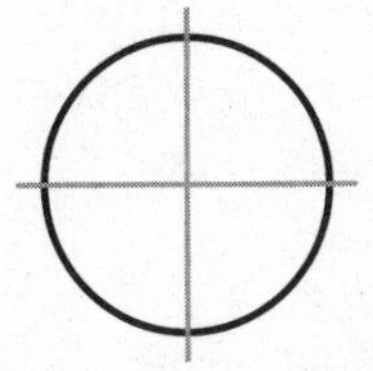

and then the other drew another two lines

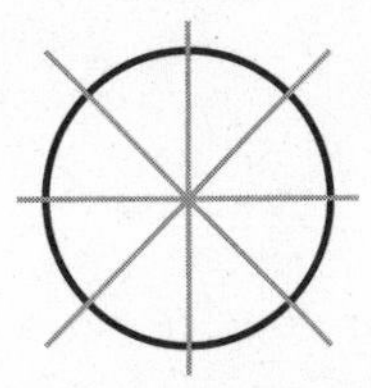

Then things got really exciting. The first child exclaimed, "There's hundreds of them!" and then the second one said, "There's *millions* of them!" and then the first one chimed in again, "You could spend your whole life drawing them and you'd never finish!" and then there was a pause, and then the other child picked up a pencil and colored in the whole circle and said, "Look, I've done it."

I was quite perplexed by that. I had to concede that they

were both right. You could spend your whole life drawing lines of symmetry on a circle and you would indeed never finish, because there are infinitely many of them. In fact, there are *uncountably* infinitely many. One way to see this is to imagine that we're specifying where the line of symmetry goes by saying what angle it makes with the horizonal.

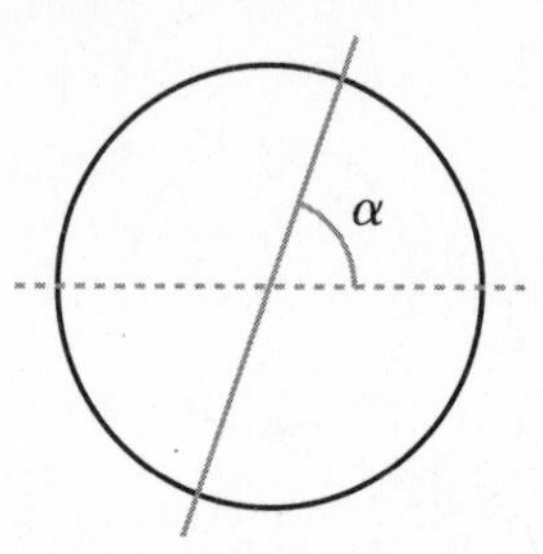

We could pick any angle from 0 to 180 degrees: or in radians, anything from 0 to π. Any bigger than that and the lines become ones we've already drawn:

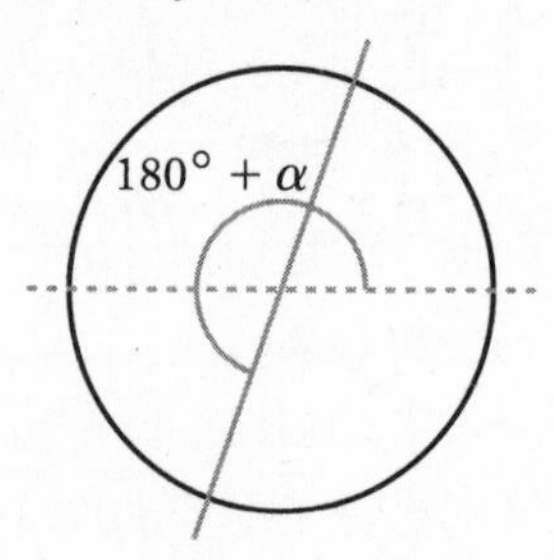

We could pick any real number between 0 and 180, not just integers or rational numbers. We've already seen that there are uncountably many real numbers just between 0 and 1, so we definitely have uncountably many between 0 and 180.

So there are uncountably many lines of symmetry on a circle, and yet if you color in the whole circle, you really have drawn them all. You might think this is cheating, because the genuine lines of symmetry would have to overlap infinitely many times at the center of the circle so really you should have infinitely many layers of pencil at the center. But even if we don't count the center and just tried to mark the places round the edge of the circle that the line of symmetry would touch, we could do it by simply drawing all the way around the edge of the circle. Have we drawn infinitely many dots in that process? Are there infinitely many dots on this line?

If there are, how wide are they? And if there are only finitely many, how many are there?

Dividing by Infinity

If we divide the line into more and more segments, the segments must get smaller and smaller. So can we divide it into infinitely many little points? This is a question of whether we can make something infinitely small by just dividing something by infinity.

Imagine you had a lottery where every natural number was a possible outcome. So there would be an infinite number of balls in the lottery machine, but each one would have a finite number written on it. In this case the chance of winning would be quite strange. In the normal UK lottery, 6 balls are drawn out of 59. There are around 45 million possible combinations of numbers, and all those combinations are equally possible. So your chance of winning is 1 in 45 million. This is a very small number (about 0.00000002) but isn't zero, although in my opinion it's so close to zero that it might as well be zero in practice. When you multiply it back by the total number of possible outcomes (45 million), you get 1, which is as it should be as it's the probability of winning if you buy every possible ticket.

In the infinite lottery there are infinitely many combinations, so your chance of winning should be "1 in infinity." What does that mean as a fraction? It can't come out to anything bigger than zero, because if it did then when we multiplied it back by the total number of possible outcomes (infinity), we would get something bigger than 1. So does that mean the probability is zero? And yet someone really could win each time. At this point you might correctly argue that no such lottery is possible in practice, but this doesn't really resolve the paradox any more than Hilbert's Hotel is resolved by saying no such hotel is possible.

This brings us back to one of our very first attempts to define infinity, when we attempted to say

$$\frac{1}{0} = \infty.$$

This produced a contradiction if we tried to multiply both sides by 0. Related to this is an attempt to say that dividing by infinity gives 0, or

$$\frac{1}{\infty} = 0.$$

Now that we know more about infinity we can immediately see some things wrong with this equation. The main problem is that the way we've defined infinity, using infinite sets of things, did not involve any notion of division by infinity. A good mathematical response to this statement would be: well, let's try it, then. Just because we didn't do it before doesn't mean it's not possible.

We have to try to investigate this the same way we did for subtraction, and go back to thinking about everything as sets of objects. This is like counting with blocks – you can't chop the blocks up (which is a source of great frustration to children sometimes). If we are thinking about the set of natural numbers, we can't cut them up into partial natural numbers.

Remember when we thought about trying to define subtraction involving infinity, we had to go back to thinking about things like $6 - 3$ as "how far do I have to count up to get from 3 to 6?" That is, we thought about solving equations like this:

$$3 + x = 6.$$

Now let's think about doing $6 \div 3$. We can think of this in two ways.

* How many times does 3 go into 6? That is, how many times do I have to add 3 to itself to make 6? This is the same as solving the equation

$$3 \times x = 6.$$

* What number goes into 6 exactly 3 times? That is, what number can I add to itself 3 times to make 6? This is the same as solving the equation

$$x \times 3 = 6.$$

The answer is 2 in both cases, because the different wording of the questions makes no difference for finite numbers. But we've seen that these are not the same question when infinity is involved.

For example, adding 3 together an infinite number of times is not the same as adding infinity together 3 times, that is,

$$3 \times \omega \neq \omega \times 3.$$

We can ask "How many times do I have to add 3 to itself to get ω?" and the answer is ω. Imagine you're the person handing out raffle tickets to the queue again. People arrive in groups of three. How many groups of three will it take before you use up one infinite book of raffle tickets? The answer is ω, because you just keep handing out three tickets at a time forever.

However, if we turn it round and say, "What number can I add to itself 3 times to make ω?" there is no possible answer. If you add 3 finite numbers together, the answer will be finite. If you add 3 infinite numbers together, each one will have to be at least as big as ω (as this is the smallest infinity), and then adding 3 of them together will be even bigger, like "forever and a day." We can think of this in terms of the raffle tickets again. If one infinite busload of people arrives, you will use up an entire book of raffle tickets (at least). Then when the next busload arrives, you're doomed to need a different color of raffle tickets.

Both of these questions were attempting to do "infinity divided by 3" but gave different answers. This shows that division, like subtraction, is not a very good concept where infinity is concerned, and that was just dividing by a small finite number. If instead we try to divide by infinity, it's even worse. Suppose we try to do $\frac{1}{\infty}$; there are two options. Either we're saying: How many times do we have to add ω to itself to make 1? This is clearly impossible as ω is much too big right from the start. Or in the other version we're saying: What number can we add to itself ω times to make 1? Again this is impossible.

Despite this, it really seems that 1 divided by infinity should be zero. Could this be a sensible answer to the above questions? If we add ω to itself 0 times, we won't get anywhere, so that makes no sense. It would be like zero infinite busloads of people arriving in your queue – you won't need any raffle tickets at all. For the second question, can we add 0 to itself ω times to make 1? This would be like zero people arriving in the queue an infinite number of times. You still won't need any raffle tickets.

At this point we could give up and say: oh well, $\frac{1}{\infty}$ is not zero, then. Or we could do something more mathematical and say: It really seems to make some sense to think in this way, so can we give it some *other* mathematical meaning that isn't based on our thoughts on infinite sets? One of the things mathematics sets out to do is take something that seems to be true according to our intuition and give a precise logical explanation to it. We shall not give up so easily.

The Opposite of Infinity

At this point you might wonder if we can just invent an infinitesimally small thing that isn't zero, as I've said we can make abstract things exist just by thinking of them. Mathematicians did try this. It vaguely makes sense, in the same way that the idea of infinity vaguely made sense until we thought about it too hard. It's a bit like an opposite of infinity. Infinity is bigger than all numbers; an infinitesimal is smaller than all numbers. When you add infinity to itself you get infinity; when you add an infinitesimal to itself you get the infinitesimal back again. And when you multiply infinity with the infinitesimal you get 1, solving the problem of the lottery probabilities.

This approach produces the same problems that just "dreaming up" infinity did in the first place. It can be done with some care (or rather, technical prowess, just as was required for coming up with a rigorous definition of infinity), but as is so often the case, it is easier, and more elegant, to sidestep the problem. If you come up to a big muddy puddle on a walk, you could step into it and hope your boots protect you, or you could step round it. (Of course, some people, especially children, enjoy deliberately stepping straight into it. This is also true in mathematics.)

Here's how we neatly sidestep the issue of dividing by infinity. Imagine you're sharing a whole chocolate cake between some people. If you're sharing it between just two people, you get quite a lot of cake each. If you're sharing it between three people, you still get quite a lot, but less than before. With four people, you get a bit less each. The more people there are, the less cake each person gets. If the numbers get really huge, it becomes silly to try and share one paltry cake. Have you ever tried sharing a cake between a hundred people? (Wedding cakes get round this by having multiple tiers, which is really multiple cakes.) What about a thousand people? A million people? At some point when there are too many people everyone will get so little cake that it will be a negligible amount, essentially none.

With a million people and only one cake, everyone will *technically* still get some cake, possibly even several billion molecules. But it will look almost like zero, and it will become more and more like zero as the number of people increases. This is how we give mathematical meaning to the notion that dividing by infinity is zero. We never actually divide by infinity (because this doesn't have a sensible meaning). Instead we go back to the idea of something approaching infinity, as in Chapter 11. We divide by something that is approaching infinity, and we find that the answer approaches zero. Some smarty-pants might come along with a microscope and say they can still see some cake, but we could always then divide it up a bit more so it had become invisible again. This doesn't mean that 1 divided by infinity *is zero*, but it gives us a way that mathematics can make sense of that intuitive idea, and is the beginnings of the entirety of modern calculus.

Zeno's Paradoxes

The ideas of calculus can be traced back a very long way. The conundrum of something being made up of infinitely many infinitely small parts was studied as many as 2500 years ago, by the Greek philosopher Zeno. Like Hilbert thousands of years later, Zeno considered some paradoxes showing that care is needed when we think about infinite things.

One of Zeno's paradoxes is basically the situation small children think of when eating delicious chocolate cake: If you eat half of what's left, and then half of what's left, and keep eating half of what's left, does it mean your chocolate cake will last forever?

Zeno put it like this: if you want to travel from A to B, first you have to travel half the distance. Then you have to travel half the remaining distance. After that you have to travel half of the new remaining distance. And you keep having to cover half the remaining distance.

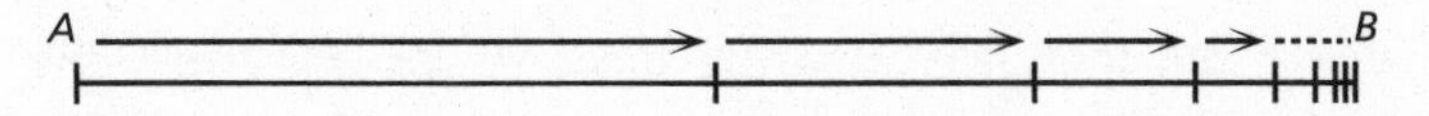

After each of these steps there will still be half of the remaining distance that you *haven't* covered yet. Surely this means you will never get there?

Mathematicians always like building up new ideas from ones they have already understood, and we can relate this situation back to the infinity of natural numbers in the following way. What we said is that we had to cover distances of a half the total, then a quarter, an eighth, a sixteenth, and so on "forever." As we said before, the natural numbers are hiding in the forever. Suppose in total we're trying to travel one mile. The stages of our journey are these:

	miles
stage 1	$\frac{1}{2}$
stage 2	$\frac{1}{4}$
stage 3	$\frac{1}{8}$
⋮	
stage n	$\frac{1}{2^n}$
⋮	

We have an infinite quantity of n's, so we also have an infinite quantity of stages in our journey. We can't finish writing out that list of how long each stage is, but we have written them all down in principle: that's what the formula with the n is for. But if we can't write down every stage of the journey, can we complete every stage of the journey? The answer must be yes, because we do in fact complete journeys, even short ones, every day. (I don't leave my house every day, but I manage to make it to the fridge several times an hour.)

A related paradox, also posed by Zeno, involves Achilles and a tortoise running a race from A to B. The tortoise is allowed a head start, say at point A_1, but goes very slowly because he is, after all, a tortoise. Now, Achilles must first reach the point where the tortoise started. By the time he gets there, the tortoise will have made some progress. Perhaps he will have made it to point A_2. Achilles must now get to this point, but by the time he gets there, the tortoise will have made a little more progress, perhaps to point A_3. Next, Achilles must now get to this point, but by the time he does, the tortoise will have gotten to A_4. Each time Achilles gets to the place the tortoise was when we last checked on the race, the tortoise will have gotten a bit farther. Does this mean the tortoise will win the race?

Both of these paradoxes involve an apparently logical argument producing a nonsensical conclusion. Clearly it is possible to make journeys and arrive at destinations. Clearly if Usain Bolt ran a race against a tortoise, Bolt would win. The point of these paradoxes is not to show that reality is wrong, but to show that something is wrong with the supposed logic in the argument.

This is a different type of paradox from the one about Hilbert's Hotel, in which the hotel could be full and yet still accommodate a new guest. In that one, the conclusion *sounds* nonsensical but only because our intuition about infinite hotels isn't very good.

The Hilbert type of paradox is called a *veridical* paradox, which means there is an entirely valid argument producing a result that seems contradictory, but isn't really. The Zeno type of paradox is called a *falsidical* paradox, in which a contradictory result is produced by an argument that seems valid, but isn't really.

In both cases the point of the paradox is to show that strange things happen when you start thinking about infinity: with Hilbert's Hotel, it's when things are infinitely *large*, and in Zeno's paradoxes, it's when things are infinitely *small*. In the Hilbert Hotel case, we have trouble getting our heads around the idea of having an infinite supply of things, because this would never happen in real life, whether it's shoes, socks, raffle tickets, or hotel rooms. However, in Zeno's paradoxes, we start to see that we do have an infinite supply of things if we allow this loophole: the things get infinitely small. They can't just *be* infinitely small, because we don't know what that means. But they can *become* infinitely small. We experience these infinite sets of things every day without ever needing to be aware of it.

Infinitely Many Infinitely Smaller Things

In the paradox about traveling from *A* to *B*, we do succeed in arriving, which means we really have covered an infinite number of distances. But this is only possible because those distances keep getting smaller, and so the time we spend on each distance also keeps getting smaller. This is in real life, not in the fantasy Hilbert world where we somehow have enough time to fill an infinite number of hotel rooms or hand out an infinite number of raffle tickets. In real life we can do an infinite number of things every day but only if the time we spend on each one becomes infinitely small.

For example, imagine you're walking a mile to the train station. Let's say you're going at a steady 4 miles per hour. This should take you 15 minutes. But what would Zeno's paradox say?

- First you have to walk the first half mile, which takes you 7.5 minutes.
- Then you have to walk the next quarter mile, which takes you 3.75 minutes.
- Then you have to walk the next eighth of a mile, which takes you 1.875 minutes.
- Then you have to walk the next sixteenth of a mile, which takes you 0.9375 minutes.
- . . .

You have to cover all these increasingly small distances, but you do them in increasingly small amounts of time. How many of these little distances are there before you reach the train station? Infinitely many – if we stopped after *finitely* many of these distances, there would always be a little bit left over.

Obviously this is an absurd way of calculating how long it would take to get to the train station, not least because at some point the tiny bit of distance left over will be smaller than the length of your foot. However, it is an important thought experiment that leads us to this realization: it seems to be possible to add up infinitely many things and get a finite answer, if the things we're adding up are getting smaller and smaller. We can't hand out an infinite number of raffle tickets, in reality, because the raffle tickets are all the same size. Even if they were getting smaller, it would still take individual, discrete chunks of time to hand each one out, so we wouldn't be able to do it. We can't take infinite mouthfuls of chocolate cake even if the mouthfuls become infinitely small, because the distance to our mouth stays the same. (Perhaps we could keep decreasing the distance to our mouth as well, and eventually end up with our face in the plate.)

There are really two puzzles here. When does it make sense to add up infinitely many tiny things? And if it does make sense, how do we tell what the answer is? This taxed mathematicians for thousands of years and was finally resolved in the nineteenth century with the formulation of calculus. We'll come to this in the next chapter.

15

When Infinity Nearly Caused Mathematics to Fall Apart (and Maybe Also Your Brain)

I have been snorkeling a few times and have very much enjoyed floating around with the fish, trying to follow the natural movement of the sea, and admiring the coral formations. Occasionally I've been gently swimming along looking at the scenery and suddenly the coral will drop off like a cliff edge. I get a strange kind of vertigo from that. It's not like standing on a cliff edge, when there's a real danger that you might fall off. I'm floating in the sea, so with the coral there's nothing for me to fall off from; it just warps my perceptions.

Certain parts of math can be perception-warping like that, and infinity is one of them. One moment you think you know what's going on, and the next moment you look in a slightly different direction and everything falls away. Mathematicians have these moments where they realize that the ground beneath their feet might not be there anymore, and then they have to hurry to fix it. Thinking about infinitely small things caused that to happen, when mathematicians realized they couldn't properly say what real numbers were.

In the previous chapter we discovered that we can do infinitely many infinitely small things in finite time, and we can fit infinitely many infinitely small things in finite space. In fact, everything is made up of infinitely many infinitely small things.

Did you feel convinced by the kinds of arguments we were making? It took mathematicians a very long time to put all this in a form that fulfilled mathematical standards of logic. It's a satisfying mathematical story of many different questions leading to the same answer, or many different clues leading to the same culprit. Zeno's paradoxes lead us to questions about adding up infinitely many infinitely small things. We'll also see how to arrive at this question through trying to understand curves by chopping them up into small straight lines, and trying to understand areas of curved shapes by chopping them up into squares.

But there's another question we rather glossed over in the first part of the book, about what irrational numbers really are. We blithely said, as is common in math classes, that they are decimals that go on forever without repeating. In fact, this is also a question of adding up infinitely many infinitely small things, because as we put more and more decimal places on a number, we're adding on tinier and tinier little fractions.

The reason we need irrational numbers in the first place is to fill in the "gaps" that are doomed to be between all the rational numbers. And it just turns out that one valid way to do it is by these infinitely long decimals, that is, by infinite sums of infinitely small things – we just have to work out what such an infinite sum means. The peculiar part of this story is that mathematicians didn't even quite realize they hadn't fixed this problem and were going round using numbers just as blithely as we do. The mathematicians Cantor and Dedekind both realized that they didn't know how to make the definitions of numbers rigorous. Dedekind was apparently preparing math classes. I know that feeling, when you're preparing to teach something and you suddenly realize there's something you thought you understood, but you don't understand it well enough to explain it to other people. In the case of Cantor and Dedekind it turned out that *nobody in the world* understood it well enough yet. Fortunately they were able to fix those foundations so the whole of mathematics did not fall apart. Rather, it accelerated forward thanks to their work on infinitely small things.

Approximating Circles

When my little nephew was about to start kindergarten, there was a list of children's math books recommended for him. I sat down and read *The Greedy Triangle* by Marilyn Burns with him. In this book the main character is a triangle who gets bored of being a triangle. He goes to the "shape-shifter" and asks for another side, so he becomes a square. Inevitably he gets bored of that and goes back and asks for another side, and he becomes a pentagon. Then he gets bored and becomes a hexagon, and then a heptagon, and then an octagon.

I got very excited reading this book because I reckoned I could see where it was going. In fact my nephew got quite excited too. (Perhaps my excitement was contagious? I hope so.) I could almost see his little brain whirring, and then he exclaimed, "I know what's going to happen! He's going to turn into a circle!"

In the end this wasn't the punch line, and I was a bit sad. Instead of him turning into a circle, which would have been an introduction to calculus, something much more morally edifying happened: he realized he had been happier the way he always was and went back to being a triangle.

My nephew's idea was the same one that was used thousands of years ago by Babylonian mathematicians to try and work out the area of a circle. It's hard to find the area of a circle because it's curved. In fact, it's hard even to say what "the area of a circle" *means*. It's somehow "the number of unit squares you can fit into it" except that you have to be able to melt the unit squares and pour them into the circular shape. Or you could do it the other way round: melt the circle and pour it into a square and see how big a square it takes up. This is what the ancient question about "squaring the circle" was. It comes up in my life (not quite daily) when I want to use a recipe for a round cake but in a square tin. I often do this because I have an adjustable square cake tin that can make any size of cake from 1 inch to 12 inches, but I only have three circular tins. Obviously I just use the formula for the area of a circle (and I usually pretend $\pi = 3$), but how did they come up with that formula in the first place?

One way you can do it without melting squares or circles (or wasting cake batter) is by approximating the circle with straight edges. The more edges you use, the better your approximation will be.

For example, if you use a square (either on the inside or the outside of the circle), the straight lines are quite far away from approximating the circle. If you try to use an 8-inch-square cake tin for a recipe that's meant for an 8-inch-round cake tin, you will not have nearly enough batter.

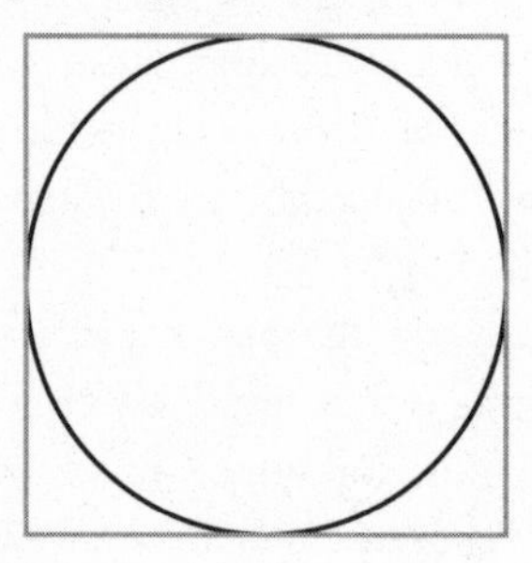

We can work out just how far off we are. Let's call the radius of the circle r. We can try this for both the inner square and outer square approximations.

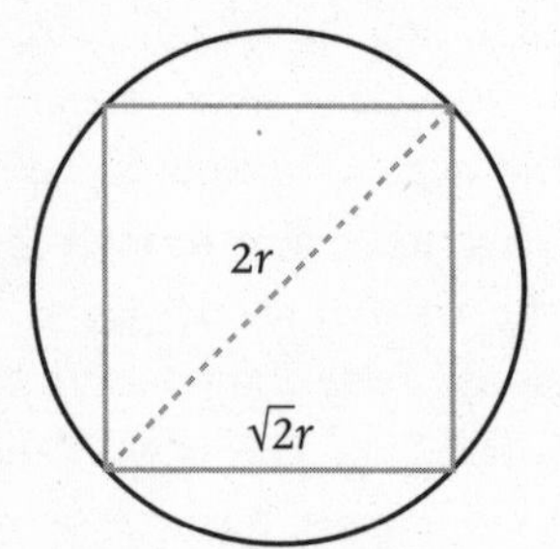

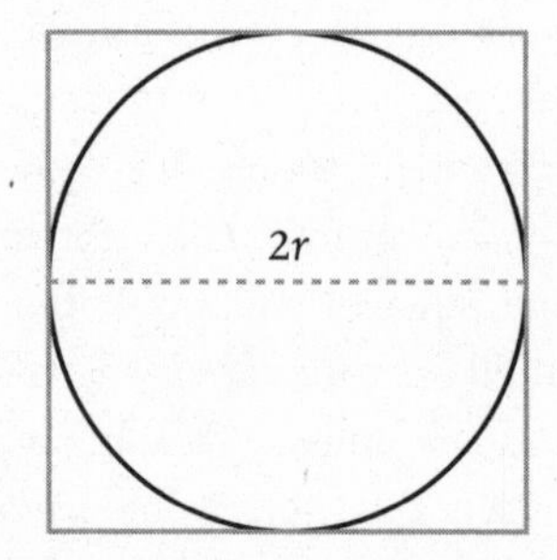

The area of the circle is πr^2. The area of the inner square is $2r^2$. The area of the outer square is $4r^2$.

You can work out the area of the inner square using Pythagoras's theorem to find the length of the edge. Or you can take the triangle shown with the dotted line and fit four of them together to make a square with the dotted line running around the outside – the length of each side is $2r$. Then this new square has area $4r^2$ and the square we want is half of that.

So we have this comparison between the "real" area and the two estimates of the area.

$$\begin{array}{llll} \text{circle} & = \pi r^2 \simeq 3.14r^2 & & \\ \text{inner square} & = 2r^2 & \text{error} \simeq & -36\% \\ \text{outer square} & = 4r^2 & \text{error} \simeq & +27\% \end{array}$$

If you use an octagon, it'll be closer.

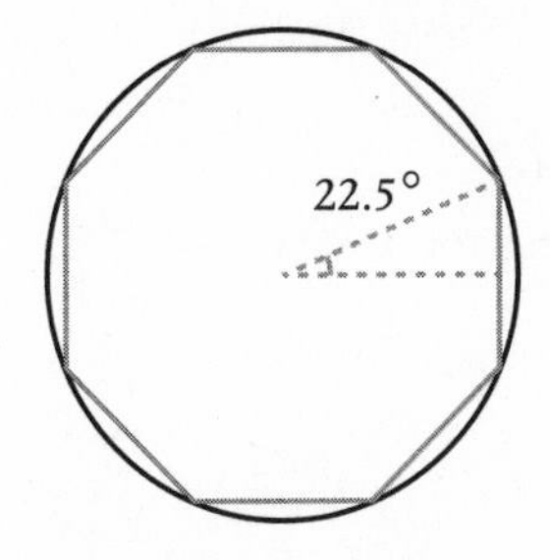

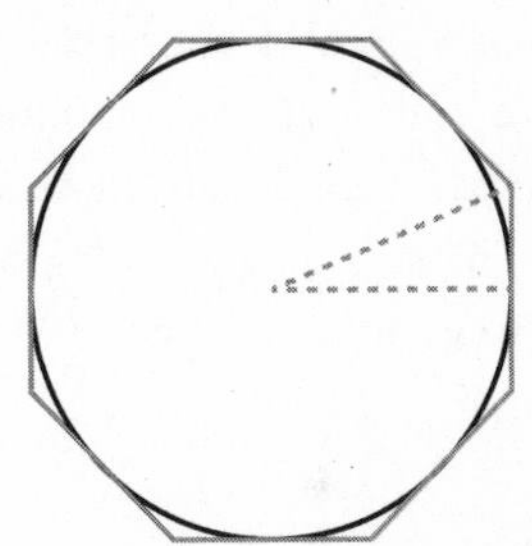

In this case we have:

$$\begin{array}{llll} \text{circle} & = \pi r^2 \simeq 3.14r^2 & & \\ \text{inner octagon} & = 2.83 & \text{error} \simeq & -5\% \\ \text{outer octagon} & = 3.31 & \text{error} \simeq & +10\% \end{array}$$

We can find the areas of these octagons by splitting them up into 16 right-angled triangles as shown. The small angle, by the center of the circle, is

$$\frac{360}{16} = 22.5$$

and we can then use a bit of traditional old trigonometry to find the lengths of the sides of the triangle. Using trigonometry might be regarded as cheating, but we're just demonstrating a point here, not simulating history.

So the idea is that you use polygons with progressively more and more sides; the more sides there are, the shorter the straight edges will be and the less error there will be between the straight edge and the real curved edge of the circle. The "true" area of the circle is sandwiched between the inner and the outer approximations, and you can get it to whatever accuracy you want by using more sides.

This is also the idea behind finding the area enclosed by a curved line, or "under" it like this:

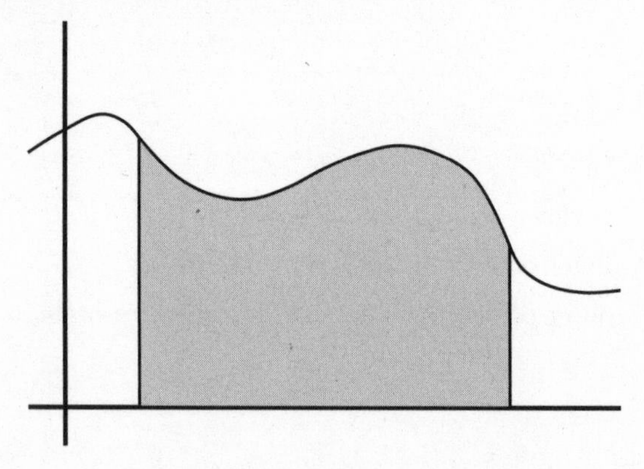

You chop it up and approximate it by pieces with straight edges, and try to make the inaccurate straight edges as small as possible. The more pieces you chop it into, the more accurate your approximation will be.

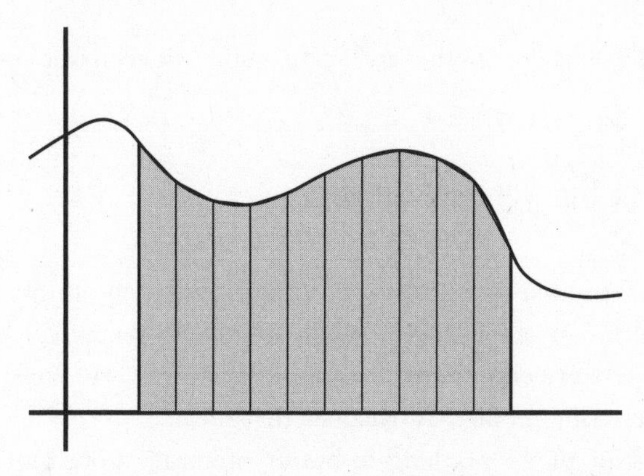

But will you ever actually get the right answer? Surely you will get closer and closer but never actually get there, because the straight edges will never precisely match the curved edge?

Getting round this conundrum is one of the big motivations behind modern calculus. The answer might seem a bit annoying at first, but is typical of how abstract mathematics works. (Admittedly, this is why math – and mathematicians – can seem annoying at times.) The answer is that we decide what "the right answer" means, in a way that suits us. This means at the very least it should not cause any contradictions, but also, it should pass basic tests of behavior so that it matches our intuition as far as our intuition goes. For example, it should give the right answer if the curve is actually a straight line!

The key to this is that in all these cases, whether it's the circle, the curve, or Zeno's paradoxes, we are doing something more and more times, and each time we do another step, we get closer to the answer. After any finite number of steps we are not at the answer, but we keep getting closer forever. We need to choose a meaning for what happens *after* we've done it forever. Our intuition has trouble here, because we are not going to live forever so we can't imagine what happens when we get there. But what we will find in this chapter is that if there is only one answer that makes sense, we might as well choose it and see what happens. Miraculously, what happens is that this fills in all the annoying gaps between the rational numbers.

Gaps in the Rational Numbers

We mentioned in Chapter 4 that if we only think about numbers that are ratios of whole numbers, there will always be gaps between them, although they're very, very tiny. Infinitely tiny, in fact. What does that mean?

First of all, it's not hard to bump into at least one gap in the rational numbers. It happens if you just draw a square whose sides have length 1. (1 something – it doesn't matter what units you use.)

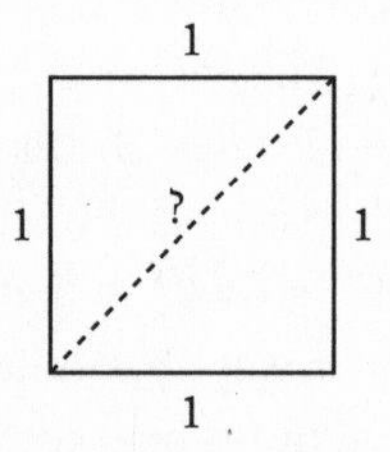

How long is the diagonal? If we draw a diagonal in this square, we have two right-angled triangles. By Pythagoras's, theorem we know that "the square on the hypotenuse is equal to the sum of the squares on the other two sides." In this case we can write d for the length of the diagonal, and this tells us

$$d^2 = 1^2 + 1^2 = 2.$$

So the length of the diagonal is some number such that when you square it you get 2. It's trying to be the square root of 2. But we can show that there is *no such number* if it has to be a ratio of two whole numbers. Here's how we do it.

Proof. We start by assuming the opposite of what we are trying to prove, so we assume that there are actually two whole numbers a and b where $\sqrt{2} = \frac{a}{b}$. The trick is also to assume that this fraction is in its lowest terms, which means you can't divide the top and bottom by something to make a simpler fraction.

Now we square both sides to get

$$2 = \frac{a^2}{b^2}$$

$$\text{so} \quad 2b^2 = a^2.$$

So far so good. Now we know that a^2 is two times something, which means it is an *even* number. This means that a has to be an even number as well, because if a were odd, then a^2 would also be odd.

What does it mean for a to be even? It means it is divisible by 2, which means that $\frac{a}{2}$ is still a whole number. Let's say

$$\frac{a}{2} = c$$

$$\text{so} \quad a = 2c$$

and now substitute that into the equation above, so we get

$$\begin{aligned} 2b^2 &= (2c)^2 \\ &= 4c^2 \\ \text{so} \quad b^2 &= 2c^2. \end{aligned}$$

Now we can do the same reasoning on b that we just did for a. We know b^2 is two times something, so it's even, which means b must be even.

Now we've discovered that a and b are both even. But right at the beginning we assumed that $\frac{a}{b}$ was a fraction *in lowest terms*, which means that a and b *can't* both be even. This is a contradiction.

So it was wrong to assume $\sqrt{2} = \frac{a}{b}$ in the first place. This means that $\sqrt{2}$ cannot be written as a fraction, so is irrational. □

So just by drawing a square as above, we unavoidably end up with an irrational number as a length. This means that there must be a little gap in the rational numbers where the square root of 2 ought to be. If we draw a circle whose radius is 1, we bump into another gap in the rational numbers, because the length of the line that we actually draw, the circumference, will be irrational (although this is much harder to prove than for the diagonal of the square).

It's a bit difficult to describe the length of the circumference of a circle, but the length of the diagonal of a square is not very contentious. If you imagine a square, the diagonal must have some length or other. If it's not a number, that's a bit of a failure of our number system. Those gaps cause other problems as well. They could, oddly, mean it is possible to draw two lines that cross each other without ever intersecting, because there might be a tiny gap there. It also relates to my favorite application of mathematics: to baby carrots.

Do Baby Carrots Exist?

I was once at a math conference munching on some baby carrots I'd bought for the coffee breaks, in an attempt not to stuff my face with cookies. I managed to get into an argument with someone about whether baby carrots are real or not. At the time I had not yet lived in the US, where, as it turns out, you can buy "baby cut" carrots. These are normal carrots that have been cut down into the size of baby carrots, just unnervingly cylindrical, regular, and rounded at the ends. I now know that these are much more widely available in the US than the type of baby carrots you get in the UK, which are genuinely carrots that haven't grown very much yet. Often they come completely intact, with the skin and the green poffley ends still attached, so they're visibly a whole carrot. Also, when I was little we grew carrots in the garden, and when we pulled them up some of them were evidently smaller than others.

These all seemed like obvious facts about carrots to me, but my American colleague was very surprised by these revelations.

However, since we were all mathematicians I was able to convince him by invoking the *intermediate value theorem*, which basically says that since carrots grow continuously from nothing, at some point they must be small.

Of course, since we were all mathematicians there was soon a refutation – how do I know that carrots grow continuously? How do I know they aren't made of green jelly until they're a certain size and then spontaneously turn into a carrot at that point, a bit like a butterfly emerging fully grown from its chrysalis? Mathematical proof can often seem like a back and forth between the mathematician and the doubter, or the mathematician and the smarty-pants. We all knew this argument was a mathematical joke, really, but this still remains my favorite application of a theorem. (From this you can correctly surmise that I am not highly motivated by grand applications of theorems.)

Anyway, what is this intermediate value theorem all about? Here's another, more mathematical way of putting it. If we draw the graph of $y = x^2$ it looks like this.

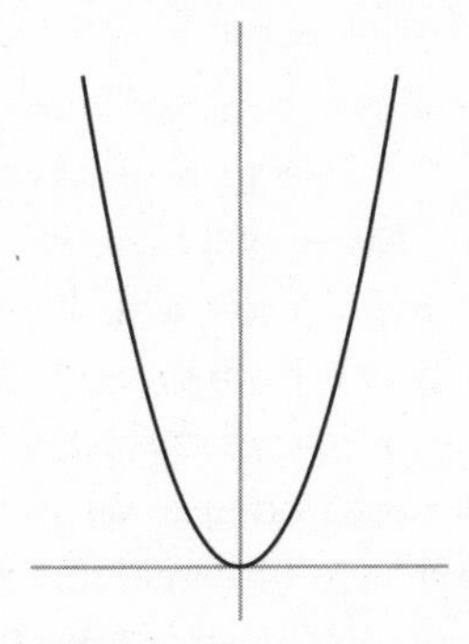

This shape is called a parabola. And if we draw the graph of $y = 2$ it looks like this.

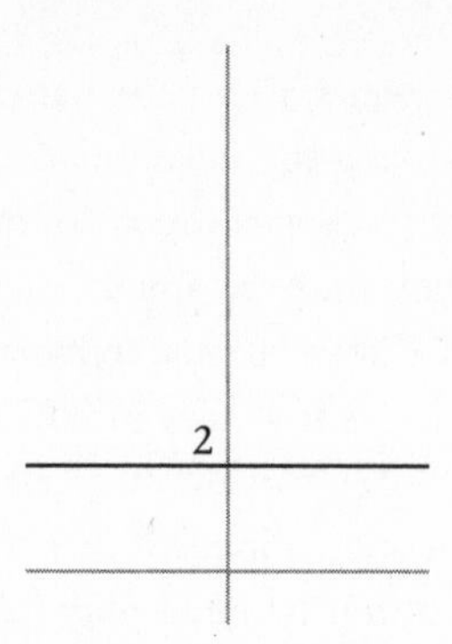

Now we can superimpose them and see where they cross:

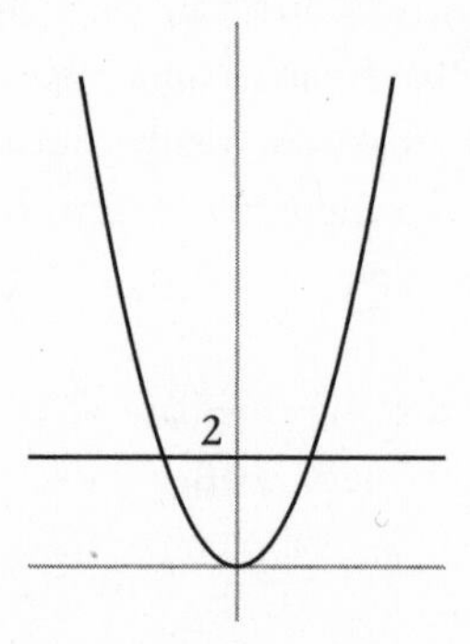

If we only have rational numbers, we know there is no rational number x such that $x^2 = 2$, which means that those two lines won't meet anywhere. There will be a tiny gap in the parabola so that the straight line doesn't actually touch the parabola anywhere, but spookily gets from one side to the other without meeting it. This reminds me of the New York City Subway, where lines cross over each other in all sorts of places without there being an interchange, whereas on the London Underground there's always (or almost always) a way of changing trains where lines cross over. (I can't find any exceptions in central London, but there are a couple way out west.)

I said the gap is in the parabola, but maybe it's in the straight line? Actually it's in both, and it's not just at that point. The gaps are everywhere, so there are all sorts of places where a straight line would cross the parabola without meeting it. Moving the line from 2 to 3 will give another place. This is one of the things that really lead us to *want* to fill those gaps in. We want to be able to know that if we draw two lines and they cross, they really actually intersect somewhere. This is the intermediate value theorem. It's called that because in its most basic form it says that if you grow continuously from height A to height B, then you have to achieve every intermediate height at some point. Anyone who is currently six feet tall must at some point in the past have been five feet tall. Carrots must at one point have been baby carrots.

Gaps Everywhere

We now know that there's at least one "gap" in the rational numbers, where $\sqrt{2}$ should be. We already mentioned in Chapter 4 that there's an irrational number between *any two* rational numbers. This means that the irrational numbers are inescapably everywhere, forcing this kind of "gap" between all the rational numbers. The real numbers are sort of stripy, between rational and irrational numbers. Of course, we have to be careful what that really means.

We are going to show that if some smarty-pants throws any two rational numbers at us, we can find an irrational number between them, no matter how close together they are. Suppose the smarty-pants has thrown us two rational numbers a and b. All we need to do is add on a tiny irrational number to a, because a rational number plus an irrational number is irrational.

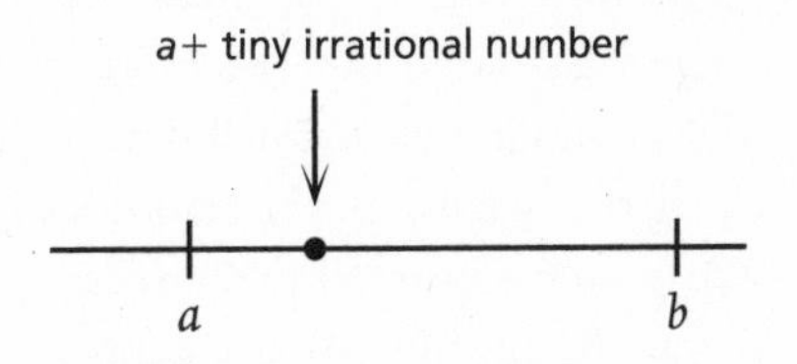

How tiny does the extra bit need to be to fall into the gap? It just needs to be smaller than $b - a$. This is easy: we can find irrational numbers as small as we want just by dividing $\sqrt{2}$ by something enormous, perhaps $\frac{\sqrt{2}}{100}$ or $\frac{\sqrt{2}}{1000000}$ or $\frac{\sqrt{2}}{10000000000}$. The number we divide by can be as big as anything, so the irrational number we get out can be as small as anything, no matter how close together our evil opponent smarty-pants makes their a and b.

> An irrational number divided by a rational number is still irrational. This can be proved by contradiction, as can the analogous fact for adding an irrational number to a rational number.

The idea here is that we can find an irrational number that's *infinitesimally close* to a. If an "infinitesimal" were a valid quantity, we might call it ε (the Greek letter epsilon) and declare that we're trying to find an irrational number that is within ε of a. However, this ε is only valid as an idea, not as a quantity, so the way we make it logically valid is to turn ε into a smarty-pants or evil opponent who can pick any real value for ε, as small as they want. The more they want to challenge you, the smaller they can pick it. Then your task is to show that no matter how small they make ε, you can still get within ε of a. This is a bit like finding a winning strategy or building a machine so that no matter what a and b some smarty-pants challenges us with, we know how to find an irrational number between them. We make a watertight strategy so we know we will never be outdone. This is the main idea behind how modern calculus is made rigorous, both when dealing with infinitely large things and infinitely small things. Infinitely large really means "so big that we can never be outdone by a smarty-pants," and similarly infinitely small really means "so small that we can never be caught out by a smarty-pants."

The tiny distance the smarty-pants is using to challenge us is usually called ε by mathematical tradition. So much so that "ε proof" is the nickname given to proofs of this kind. The idea of the ε is that it represents an arbitrarily small number that our smarty-pants is going to throw at us. It would be bad strategy for them to throw a large number at us after all, because it would make it easy for us to win. It harks back to the time when mathematicians tried to make this sort of argument precise by using actually "infinitesimally small" quantities. However, ε is just potentially small, not a fixed small number. So ε is sometimes used informally as synonymous for "very tiny amount." Mathematicians will say things like "I feel like I've been ε away from finishing this paper for months." In fact, there's a mathematical joke that simply goes like this: let ε be a large negative number. If you've spent years slaving away over ε proofs, this can be enough to cause a fit of nervous hysterics.

This type of justification is quite different from the kind where you solve equations or do geometrical constructions. It feels like you're skirting around an issue, or never quite getting there. This is a valid feeling, and often leaves students of calculus feeling very disconcerted when they first see it: it's not how they're expecting mathematics to be. The first time you see something called calculus, it's usually a quite well-organized set of rules and procedures for calculating things like slopes of graphs and areas under curves. The question of why these give the right answer gets pushed under the carpet because it's too mind-boggling. The reason it's mind-boggling is because you have to get your head around the notion of things being infinitely small. We've already seen how infinitely large things make our minds boggle, and now it's time for some infinitely small things.

Imagine that someone is trying to challenge you to shoot a target from some distance away. You're allowed a few misses at the beginning, but in order to pass the test you have to be able to keep hitting the target indefinitely after a certain point. If you pass this test, the opponent makes the target smaller and you have to try again. And if you pass that test, they're going to make it even smaller. And they're going to keep making it smaller to try to catch you out. You only count as being able to hit a target if you can pass this strenuous test *no matter how small* they make the target.

Target Practice

Let's try this "target challenge" for an example of something that is supposed to get infinitesimally small: $\frac{1}{n}$, as n gets bigger and bigger. We've seen that this is supposed to be like "dividing by infinity, giving zero," and now we can make some sense of it. The idea is that $\frac{1}{n}$ will never actually equal zero, but if 0 is the bull's-eye of our target, we will always be able to keep hitting the target no matter how small our evil opponent makes it. It's not enough just to hit the target once. What we have to do is allow ourselves a few misses at the beginning and then be able to keep hitting it indefinitely. You might be wondering how many misses is "a few." The answer is: any finite number. Because when we're talking about "forever," a finite number of misses at the beginning isn't going to count for much by comparison.

For example, suppose the evil opponent makes the target only 0.01 wide (as a radius). So the range we're trying to hit is this.

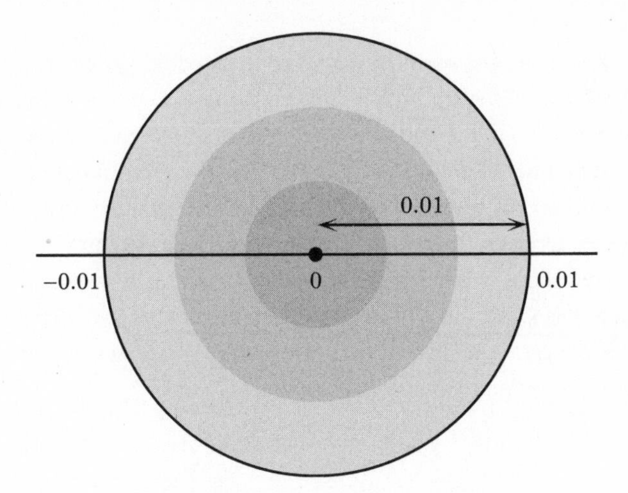

Our target should only be a portion of the real line, but I've drawn it to look like a circular target so that it looks more like target practice.

Our attempts at hitting the target consist of making n get progressively bigger, so they go like this:

$$1, \frac{1}{2}, \frac{1}{3}, \frac{1}{4}, \frac{1}{5}, \frac{1}{6}, \frac{1}{7}, \frac{1}{8}, \frac{1}{9}, \ldots$$

Our tenth attempt is $\frac{1}{10}$, which is 0.01. This one will sort of ricochet off the side of the target. Our first one that will definitely hit the target is the next one, $\frac{1}{11}$. From this point on we're fine – all subsequent attempts will be smaller than that, so they will all hit the target.

We have passed this stage of the test. But the evil opponent is undaunted. They replace the old target with a new, smaller one that's only 0.001 wide. This time it will take us much longer to hit the target, but that is allowed – remember we're allowed any finite number of misses. Now the one that will ricochet off the edge of the target is $\frac{1}{1000}$, as that is exactly 0.001. But the next one is $\frac{1}{1001}$, which is smaller so will hit the target. All subsequent attempts will be even smaller so will all hit the target, and so we pass the test.

Now it might seem that this process could go on forever, but we can come up with a clever mechanism to show that *no matter how small* the evil opponent makes the target, we can still win. It's a bit like how we showed that there would definitely be an irrational number between a and b, no matter how small the gap between them was. This time, we want to show that we can pass the test on a target of size ε no matter how small ε is. Again, ε isn't a fixed infinitesimally small size, because that can't be made rigorous. Instead it's a potentially small size, as small as the evil opponent smarty-pants wants.

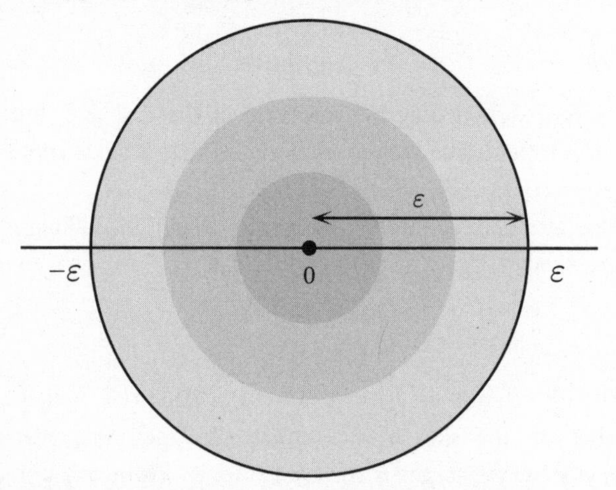

It's just a matter of finding a time beyond which all of our $\frac{1}{n}$ will hit the target. Now, because the $\frac{1}{n}$ keep getting smaller as we go on, we just have to find any old one that hits the target, and we know that from there on we will be fine. It doesn't matter if we find the first one or not, we just have to find one.

What this means is we have to find n such that $\frac{1}{n} < \varepsilon$, which we can definitely do by making n really huge, as huge as necessary. This is just like when we made a tiny irrational number by dividing $\sqrt{2}$ by something as huge as necessary.

If we want to be very precise, we can work out how huge a number we have to divide by to make something small enough. In this case we are trying to get $\frac{1}{n} < \varepsilon$, and if we rearrange this, it becomes $\frac{1}{\varepsilon} < n$. So we just have to find an n that's bigger than $\frac{1}{\varepsilon}$, whatever that number happens to be. And we know we can always do this, because the n's get bigger and bigger and never run out.

This means that in principle we know we will always be able to pass the target test, no matter how small a target the evil opponent chooses.

Mathematically we say that 0 is the *limit* of this sequence:

$$1, \frac{1}{2}, \frac{1}{3}, \frac{1}{4}, \frac{1}{5}, \frac{1}{6}, \frac{1}{7}, \ldots$$

This doesn't mean the sequence of numbers will ever get to 0, but it means that it will keep getting closer and closer, with the precise meaning of the target challenge we just described.

Cake Again

We can now use this method to give a rigorous meaning to the question of whether we can make our cake last forever by eating half of the cake, and then half of the rest (which is a quarter), and then half of the rest, and then half of the rest .... forever. You might think your cake will last forever, but the bit that's left will eventually get so small as to become negligible, so you will basically have eaten the whole cake.

This can be pictured in a very satisfying way on a square cake, as long as you don't mind that some pieces will be square and others will be oblong.

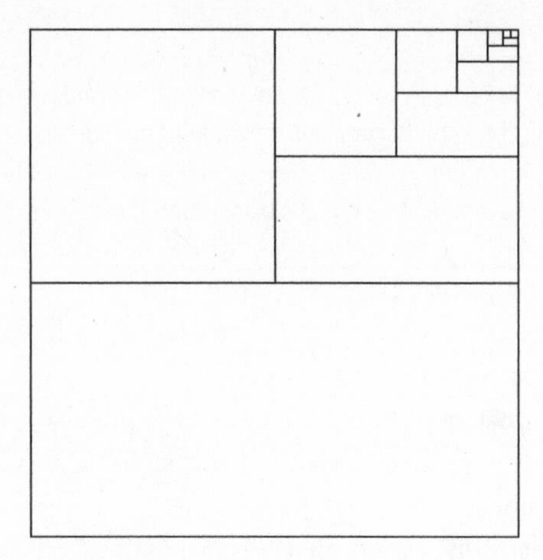

The question is, how much cake will we have eaten after we get to forever? At each stage we eat these amounts:

$$\frac{1}{2}, \frac{1}{4}, \frac{1}{8}, \ldots$$

except this time it's not the amount we eat at each stage that we're interested in, it's the *total amount* we're eating. The total amount we've eaten after each stage is

$$\frac{1}{2}$$

$$\frac{1}{2} + \frac{1}{4}$$

$$\frac{1}{2} + \frac{1}{4} + \frac{1}{8}$$

$$\frac{1}{2} + \frac{1}{4} + \frac{1}{8} + \frac{1}{16}$$

$$\vdots$$

Now we're going to claim that if we "keep going forever," we'll eat all the cake. By this we mean that we'll get closer and closer to eating the whole cake, as close as any smarty-pants could ever challenge, according to the target-test method from before.

Mathematically speaking, we're saying that the *limit* of the amount of cake we'll eat is 1, the whole cake.

So along comes the smarty-pants to challenge us. This time the bull's-eye is 1. They start by giving us a target that is 0.1 wide (measured by radius).

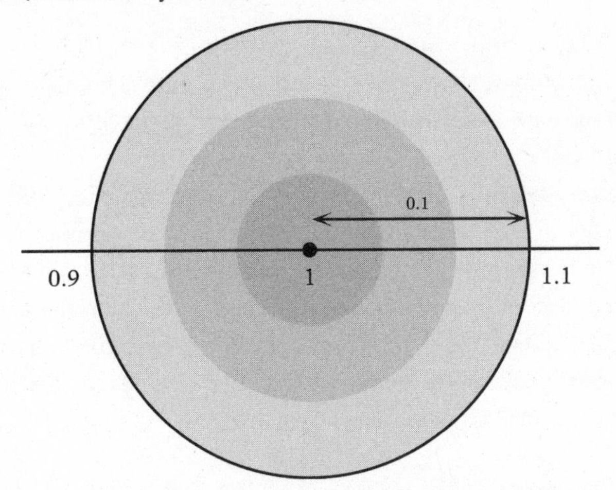

We have to show that after a finite number of allowed misses we can hit this target and keep hitting it with all subsequent attempts. This time the subsequent attempts are the successive rows of the above table. You might think we need to add those fractions up to find the answer at each stage, but we can be much more lazy than that (and one key to math is to be as lazy as possible). It's much easier to think about how much cake is *left*, and that will tell us how far away from the bull's-eye we are at each stage.

After the first attempt, there's still $\frac{1}{2}$ the cake remaining, so we're $\frac{1}{2}$ away from the bullseye of 1. Seeing as the target is only 0.1 wide and $\frac{1}{2} = 0.5 > 0.1$, this has not hit the target.

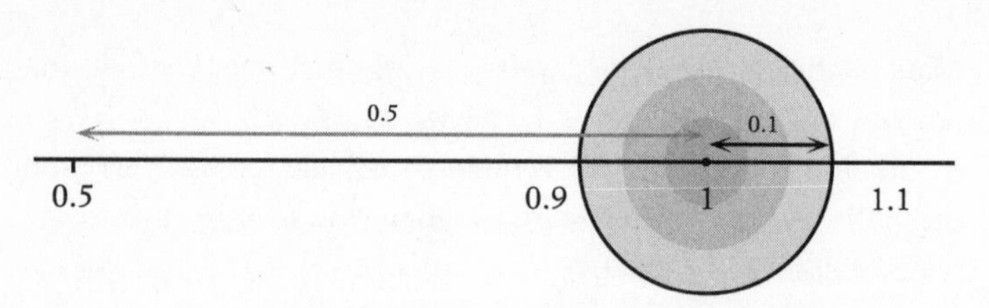

Our next attempt gets us closer, but we have still left $\frac{1}{4}$ of the cake, meaning that we are $\frac{1}{4} = 0.25$ away from the bull's-eye of 1. Now 0.25 is still bigger than 0.1, so we have not hit the target.

After the next attempt there is 0.125 remaining, still not close enough to be on the target. After the next attempt there is 0.0625 remaining. Finally this is smaller than 0.1, so this has landed close enough to the bull's-eye of 1 to be on the target. All subsequent stages will take us even closer to the bull's-eye, so we will still be on the target forever. So we have passed the target test for this particular target.

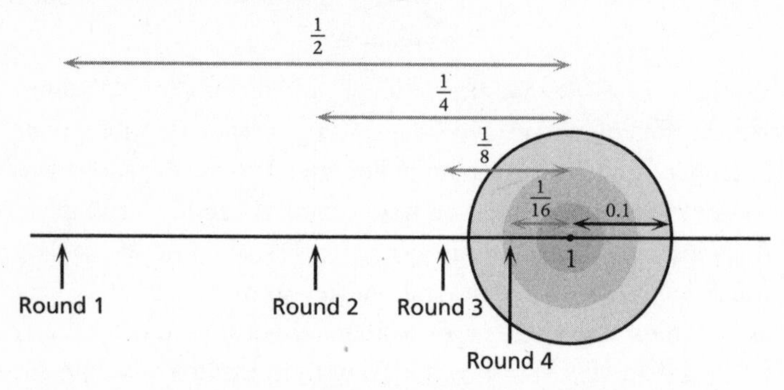

In fact, we can always pass the test, no matter what size of target we're given. Suppose the smarty-pants picks a target of width ε. It's a bit harder now because we need to make a winning strategy that will work no matter what ε is. When the size of the target was a specific number (like 0.1 above), we could just keep testing to see how much cake was left at each stage and stop when it was less than that number. To make a strategy that will work for *any* ε, we have to work out when the leftover cake will be less than ε without knowing what ε is. This is where math can start seeming impossible to get your head around, because instead of an actual number, it has become a letter, and you don't know what it is.

What we have to do is realize that after each stage of eating, we are halving the amount of cake that's left over, so the amount left goes

$$\frac{1}{2}, \left(\frac{1}{2}\right)^2, \left(\frac{1}{2}\right)^3, \left(\frac{1}{2}\right)^4, \ldots$$

and so the amount left after n stages of eating is

$$\left(\frac{1}{2}\right)^n = \frac{1}{2^n}.$$

Informally, we can now say "as n gets bigger and bigger, 2^n becomes *enormous*, so by making n bigger we can always make the leftover smaller than any ε."

To be more precise, we need to get to a point where this leftover is less than ε, that is,

$$\frac{1}{2^n} < \varepsilon$$

which is the same as saying

$$\frac{1}{\varepsilon} < 2^n.$$

Now, 2^n is always bigger than n (as long as n is bigger than 2), so if we just make sure $n > \frac{1}{\varepsilon}$, then we'll get this string of inequalities:

$$2^n > n > \frac{1}{\varepsilon}$$

giving the result we need. This is not the most efficient way to do it, in the sense that we haven't found the first moment we hit the target. But that doesn't matter. We just need to know that we *can* hit it forever; we don't need to know exactly when this happens. This can feel satisfying if you like being lazy, or unsatisfying if you like precision. I like both, but laziness wins.

So we have a watertight strategy for passing the target test for any size of target. This means that by the mathematical definitions, the limit of the amount of leftover cake is 0, so the limit of the amount we eat really is 1. This is the meaning we have given to "what happens if we keep going forever."

Note that this "limit" doesn't mean a boundary. It doesn't mean this is the most cake that we can eat (although as it happens, this *is* the most cake we can eat since we only started off with one cake). The use of this word "limit" is supposed to capture the idea of something going on forever and eventually settling down and not moving around anymore.

This explanation is also how we extract ourselves from the conundrum of Zeno's paradox about traveling from A to B: we have to travel half the distance, then half the remaining distance, and so on. The key is to look at *how long* it takes us to do each stage of the journey. If we keep going at the same speed throughout, then each stage will take half the time of the previous stage. So although we have to go through an infinite number of stages, they get so short that the total time we spend doing it is still finite. In fact, it's basically the same calculation as the chocolate cake example. If the first stage takes us half an hour, the second stage takes a quarter of an hour, the third takes an eighth (not that we usually measure hours in eighths), and so on. If we keep adding up these times indefinitely, we have a "smarty-pants target test" situation, and it's just like the answer to the chocolate cake: the whole thing adds up to an hour, despite there being infinitely many stages.

So it turns out we can do an infinite number of things in a finite space of time, and we do it every single day by going from place to place. Unless we stay in bed all day, but that's beside the point.

Recurring Decimals

We can now give the target-test treatment to recurring decimals as well. Let's think about that most confounding one, $0.99999999\cdots$, which is also written as $0.\dot{9}$, and you might have been told at some point in your life that this "equals 1." What does this actually mean? We tend to think of this recurring decimal as "$0.999\cdots$ with the 9's going on forever," but now we know how to deal with it in a rigorous mathematical way. It's just like eating the cake. We gradually proceed down the decimal places. After the first decimal place we have 0.9. After the second decimal place we have 0.99. After the third we have 0.999. And so on. The definition of $0.\dot{9}$ is that it is the *limit* of this process. So when we are saying $0.\dot{9} = 1$ all we are claiming is that the *limit* of this process is 1. It doesn't mean we actually ever reach 1. It means that if an evil opponent comes along and sets up a target for us with a bull's-eye at 1, we can pass the target test no matter how small they make the target.

Let's try it. Let's say the first target is 0.01 wide. So our target looks like this:

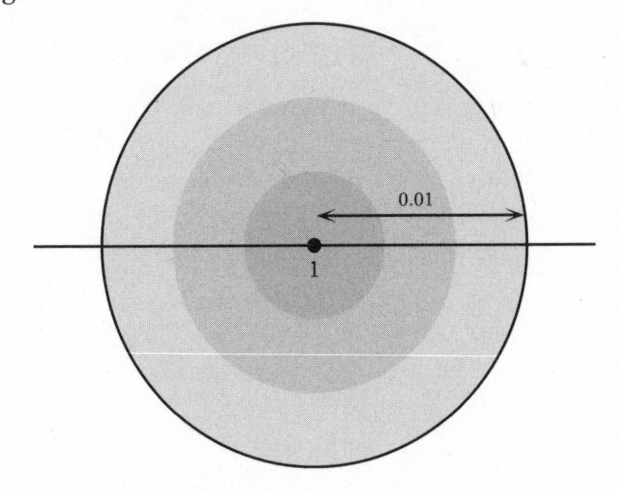

Our first attempt is 0.9. How far away is this from the bull's-eye 1? It is 0.1 away, which is bigger than 0.01. So we have missed the target.

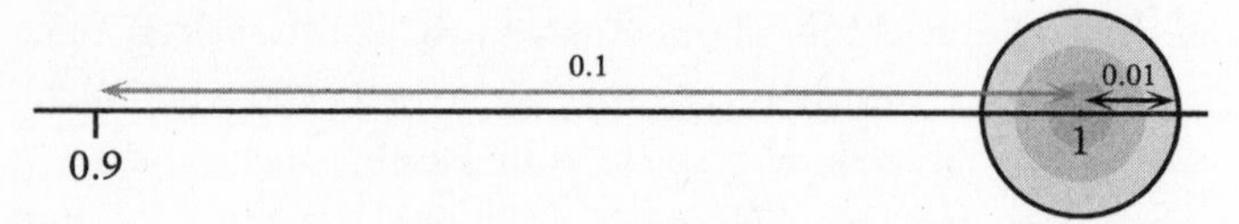

Our next attempt is 0.99. This is 0.01 away from the bull's-eye 1, which means we have ricocheted off the edge of the target.

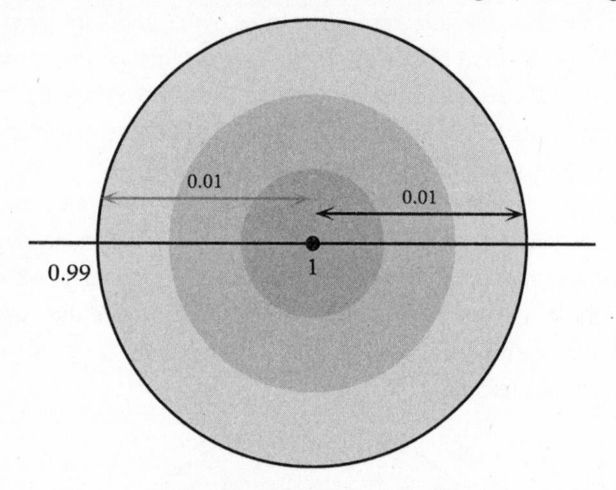

Our next attempt is 0.999. This is now only 0.001 away from the bull's-eye 1, which means we have definitely hit the target this time. Morever, all our subsequent attempts get *closer* to the bull's-eye, so we know that we will now hit the target with all future attempts. So we have passed the target test.

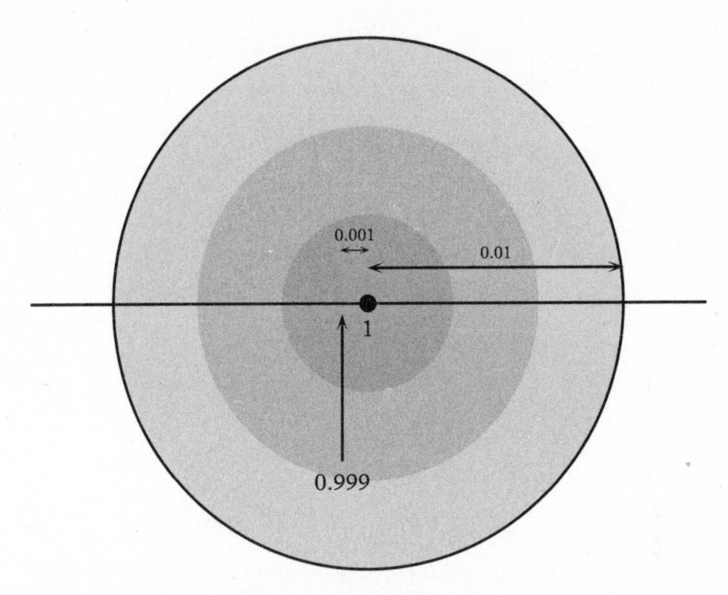

Now we have to be sure we can pass this test no matter how small the target is. Let's suppose that the evil opponent gave us a target that was ε wide. When will we start hitting this target?

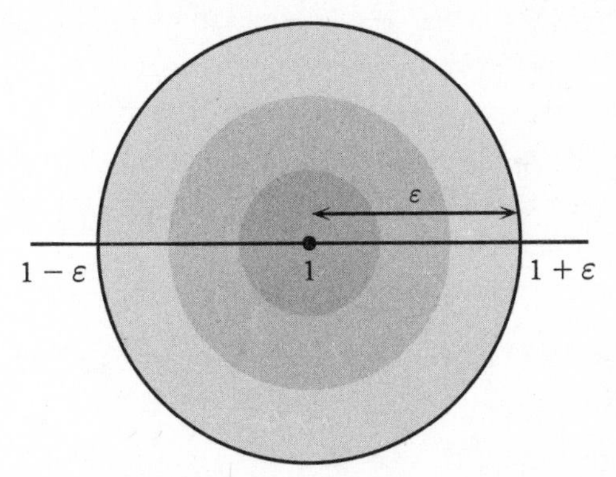

We know we will hit it once our error margin is less than ε, which means we need $0.0000 \cdots 001 < \varepsilon$ for some number of 0's. This is definitely possible because we can put in as many 0's as we want.

This is pretty much the same as the cake example, except that instead of halving the leftover part each time, we are dividing it by 10. So we need n such that

$$\frac{1}{10^n} < \varepsilon$$

which is the same as

$$\frac{1}{\varepsilon} < 10^n.$$

Similarly to the last time, we can use the fact that $10^n > n$, so if we just pick $n > \frac{1}{\varepsilon}$ we then have

$$10^n > n > \frac{1}{\varepsilon}$$

as we wanted.

So if you think that $0.999\cdots$ will never "actually" hit 1, you're right but missing the point. The definition of this recurring decimal already has that fact built into it, so it never actually has to "hit 1" in order for the recurring decimal to equal 1.

Another way to think of this is to ask: If $0.999\cdots$ isn't 1, what is it? It can't be less than 1, because we've just shown that it gets closer to 1 than any possible size of gap. So however small you think the gap is between $0.\dot{9}$ and 1, the actual gap can't be any bigger than 0. We now have two options.

1. Declare that $0.999\cdots$ makes no sense and isn't a real number.
2. Declare that $0.999\cdots$ is 1.

Let's compare these options. If we take the first option, we will not be able to make sense of any other infinite decimal expansions either, and so we will not get to fill in the gaps in the rational numbers. If we take the second option, we get to fill in the gaps in the rational numbers and nothing logically bad happens at all, we just feel a bit queasy. If you remember that

$0.999\cdots$ is just shorthand for "the limit as n approaches infinity of $0.99\cdots9$ with n decimal places," then I hope your queasiness might subside and you will be able to enjoy the ride.

> The cake example corresponds to a recurring fraction if we use binary fractions instead of decimals. This is because $\frac{1}{2}$ in binary is 0.1, and then if we add on $\frac{1}{4}$ we get 0.11, and then if we add on $\frac{1}{8}$ we get 0.111 and so on, so the amount of cake we eat if we go on forever is the recurring binary fraction $0.1111\cdots$. And by an argument as above, we can show that this limit is 1.

Aside on "Forever"

We've seen a lot of situations where we're trying to imagine what happens when something goes on forever, whether it's handing out raffle tickets, guests arriving at a hotel, floors of a skyscraper. This is hard because none of these forevers is possible in reality. How can we imagine what would happen if something impossible became possible?

It is a fun device in fiction to change one thing in reality but leave everything else the same. For example, Superman lands on Earth but everyone else on Earth is a normal human being. Or someone has built a time-traveling machine but all the humans are still normal human beings. This is the sort of thing we're trying to do in our "forever" thought experiments as well.

Mathematically and logically this is difficult, because logically if you make something false become true, then *everything* becomes true. It's a bit like the fact that if you make $1 = 0$, then *everything* is doomed to be equal to 0.

Think about an argument where someone says, "If you're right, then I'm the Queen of Sheba." What they're saying is that if you're right, then everything might as well be true, including something entirely fanciful like me being the Queen of Sheba.

In logic, you can't just take something that's false and make it true and think that there will be no consequences – the consequence is that your logical system won't be consistent anymore. The only way to do it that will keep the logical system consistent will be to make *everything* true. "True" and "false" end up meaning the same thing, and the world implodes. That's why imagining Hilbert's Hotel and infinite cookies are only thought experiments, not logical arguments. That's why they're open to debate, and why we have to work hard to turn them into genuine mathematical arguments that are no longer open to debate, at least not by people who understand mathematics.

Other Long Decimal Expansions

We now have a hint of what an "infinitely long decimal expansion" means. It means this thing about targets. Every time you add a decimal place to a number, you're changing it a little bit (unless the new decimal place is 0), but as you add more and more decimal places, the amount that you're changing the number gets smaller and smaller. At some point it becomes very negligible; that is, you know you will keep hitting the evil opponent's target, no matter how small they make it.

With recurring decimals, we extend the decimal places according to a repeating pattern. The pattern doesn't have to start at the beginning, and the repeating cycles can be as long as you like, but eventually they repeat. Here are a few examples:

$$0.111111\cdots \qquad 0.\dot{1}$$

$$0.131313\cdots \qquad 0.\dot{1}\dot{3}$$

$$0.18640278278278\cdots \qquad 0.18640\dot{2}7\dot{8}$$

The right-hand column is the more unambiguous way of

writing down the decimals so that we're all completely clear which part is going to repeat.

The strange conundrum with irrational numbers is that there is no pattern to the way that we're extending the decimal places. The digits are random and never repeat themselves, not even in very, very long cycles. So how can we tell what bull's-eye this number is heading toward? This is a good question, and the answer is: we can't.

This is the point where we flip everything upside down and do it all backward, which is what happens sometimes in math and might make you feel queasy. Being on a boat (especially a small boat) can make me feel very queasy unless I tune into the way the boat is rocking with the waves and fully let myself be rocked with it. At that point it becomes positively exciting. (However, I've never managed to do this with those big swing rides at theme parks. Those just make me feel ill.) What I find much harder is coming back on land again. I've usually gotten so tuned into the rocking of the boat that I can't tune into the boring fixed land afterward.

Here's how we're going to turn ourselves upside down with the targets. Previously our evil opponent placed a target around the bull's-eye that we were aiming for, and we had to pass various tests. This time *we don't know where the bull's-eye should be*. We just have a target that is getting smaller and smaller. As long as we can keep hitting the target in the same way as before, we win. What do we win? We win the knowledge that there was a bull's-eye somewhere, even though we don't know where it was.

This is what happens with infinitely long decimal expansions of irrational numbers. We know that as the expansion gets longer the number will settle down and keep hitting a very small target, but we never get to find out where the bull's-eye is. For example, let's think about π. It's a remarkable fact that no matter what circle you take, the ratio of its circumference to its diameter will always be the same. It's the same ratio for all different sizes of circle. This ratio is an irrational number and we call it π.

But what number is π? We don't know. We have very long decimal expansions of it (say, to 10 trillion digits), but this doesn't mean we know what the entire decimal expansion is. We can even compute any given digit without computing the previous ones, by an algorithm of Simon Plouffe from 1995. But we still can't know the entire decimal expansion at once.

We know that we can hit any infinitesimally small target by going very, very far down the decimal expansion, because after, say, 5 million digits the number is hardly changing at all when we add more digits. But it is still changing, and we do not know what bull's-eye it is heading toward. This is the strange fact about the irrational numbers. They are everywhere, and we really can't say what they are, except by some circuitous arguments using shrinking targets. It just so happens that π can be *characterized* precisely without any reference to decimals, because it is simply the ratio of any circle's circumference to its diameter. Likewise, $\sqrt{2}$ can be characterized as the positive number that squares to 2. However, most irrational numbers can't be characterized in this way. So how do we say what they are? It's difficult.

What Are the Real Numbers?

No wonder it took so long to pin down the real numbers. It's quite easy to say what the previous types of number are up until that point – we did the natural numbers, the integers, and the rational numbers without too much difficulty. Once you start letting irrational numbers in, the situation becomes very difficult.

In one of those exciting moments in the history of mathematics, two people worked out how to do it, in different countries, in different ways, at almost exactly the same time. It's remarkable how often this happens. It's almost as if there's something in the ether that makes mathematicians in different places around the world able to solve the same problem at the same time. Not that there is an ether. Perhaps it's just that mathematical research reaches a certain point leading up to it and then the time is ripe for it. The real numbers were finally properly pinned down by Cantor and Dedekind, in completely different ways, independently, in 1872.

Cantor did it using the shrinking-target method that we've just been describing; confusingly the name of another mathematician, Cauchy, is usually associated with this construction because the shrinking-target method is based on his ideas. It was Cantor who put the method to use for constructing the real numbers. Dedekind did it by something more like finding the gaps between the rational numbers. Both of these methods end up with real numbers that look extraordinarily unlike anything we're used to thinking of as a number. This is just like our previous construction of infinity, which turned out to be the "quantity of numbers in a set" even though we don't know what the set is.

Using Cantor's method, a real number is constructed as a sequence of rational numbers that meets the target-practice condition. Here are a couple of famous examples. The irrational number e can be defined via the sequence of rational numbers: $\frac{1}{n!}$. Remember that $n!$ is

$$n \times (n-1) \times (n-2) \times \cdots \times 3 \times 2 \times 1.$$

Here, as a sort of technicality, we start n at 0 and define $0!$ to be 1.

One justification for this is that $n!$ is the number of different orders in which you can play the n songs on your iPod. If you have 0 songs on your iPod (as I currently do on my new phone), there's only one way in which you can play them, and that is by doing nothing.

The numbers we're considering here are:

$$\frac{1}{0!}, \frac{1}{1!}, \frac{1}{2!}, \frac{1}{3!}, \frac{1}{4!}, \frac{1}{5!}, \ldots$$

and we're going to add them all up forever. This wins in our evil opponent's target test, and so although we don't really know what the bull's-eye is, we know it's there somewhere, and we call it *e*.

Miraculously, this is the same number you get by a more well-known definition of e, which is via the function e^x. This is a function satisfying the following two properties:

1. The function's gradient everywhere is itself. That is, if you plot a graph of it, the slope at any point is just the *y* value of that point.
2. The *y* value (and gradient) at 0 is 1.

The number e is then the *y* value (and gradient) at 1.

Here's another example. This time we add up these rational numbers:

$$4, \ -\frac{4}{3}, \ \frac{4}{5}, \ -\frac{4}{7}, \ \frac{4}{9}, \ -\frac{4}{11}, \ldots$$

This also beats our evil opponent's target test, and again we don't really know what the bull's-eye is. We call it π. The fact that this is π is a curious result that can be proved using some quite sophisticated techniques of calculus.

Showing that these complicated abstract objects behave like numbers takes a lot of work. We have to show how to add them up and how to multiply them, and prove the rules that we want to be true about numbers, like that the order of addition and multiplication doesn't matter, that we can subtract from both sides of an equation, and so on. It is amazing that something we can understand so instinctively can be so difficult to turn into rigorous mathematics. This can make mathematics seem frustrating to some people, impotent to others, and pointless to yet others. To me it is simply fascinating that our gut instincts can be so strong, and so difficult to understand with our brains. It doesn't mean we shouldn't try, although we can also be glad that great mathematicians like Cantor and Dedekind did it for us, so that we can marvel at their solutions and not have to try and do it ourselves.

16 Weirdness

Before we end, here are some weird facts that come out of our new understanding of the infinitely small. Infinite things and finite things start getting mixed up in strange ways.

First of all, here's how to make an infinite number of cookies with a finite amount of cookie dough. You make the first cookie, and then you make the second one with half as much dough as the previous one.

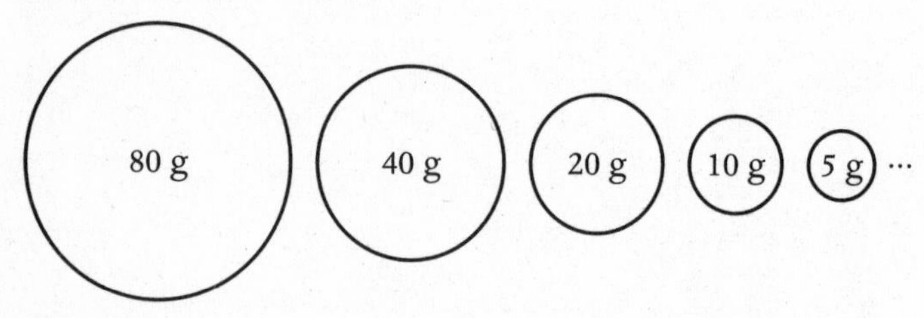

You make the next one with half of that, and then half of that, and so on. As long as the first cookie wasn't a giant monster cookie that used up half the dough (or more), then you will never run out of dough. (Making a single cookie with all your dough and baking it in a pizza dish seems to be something people do these days.) You will have an infinite number of cookies; the only snag will be that they will also be infinitely small and once they're smaller than a certain size you won't be able to see them anymore.

How about making the cookies get small just a little bit more slowly? Instead of making each cookie with half the dough of the previous one, you could go back to the first cookie and just take a slightly smaller proportion of that dough. So the second cookie could be half the first one, the next one could be a third, the next one a quarter, and then a fifth, and so on.

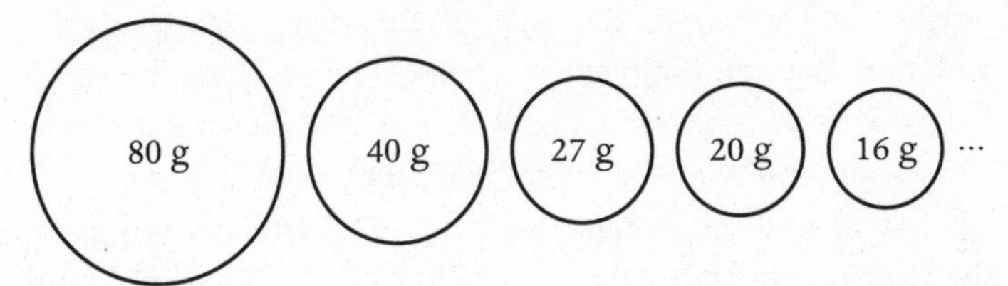

The trouble is this time you will need an infinite quantity of dough.

On the other hand, you could make the second cookie with half the *radius* of the first, and the next with a third the radius, and then a quarter the radius, and then a fifth, and so on.

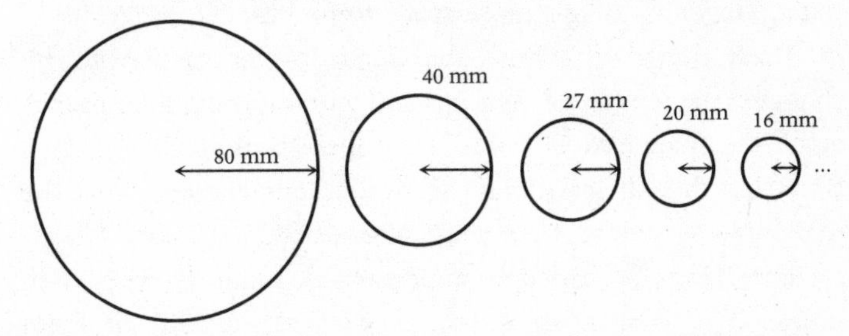

This time you will be able to stick within a finite quantity of dough. However, if you line all these cookies up in a long line, they will stretch an infinite distance.

What?

The Harmonic Series

The progression of numbers that we're thinking about is

$$\frac{1}{2}, \frac{1}{3}, \frac{1}{4}, \frac{1}{5}, \frac{1}{6}, \frac{1}{7}, \ldots$$

and is called the *harmonic series*. It is related to harmonics in music – these are the wavelengths that give the harmonics of a note, expressed as ratios of the main note. For example, on a violin if you play the G string without putting any fingers down, you get the note G. You can play higher notes by putting fingers down and shortening the part of the string that's actually vibrating. But you can play "harmonics" by touching the string lightly at certain places, without completely pressing it down. They sound more ethereal and floaty than normal notes.

If you touch the string halfway up, the harmonic you get is the G an octave higher than the main G. If you touch it a third of the way up, you get the D an octave and a fifth above the original G string. You get the next harmonics by touching the string a quarter of the way up, and then a fifth, and then a sixth. You can't hear much more than that on a violin string, although you can hear some more on a cello string, as the strings are much longer so there's more space to find the harmonics.

These are the same notes that a brass player can play just by changing the tension of their lips and without using their fingers at all. The main note of a brass instrument is more likely to be a B-flat, but then the harmonics are at the same *intervals* above the main note: an octave, then a fifth above that, and so on.

The harmonic series is important in math as well as in music. The weird fact is that if we keep adding all these fractions up "forever," it *never* hits any kind of target. That is to say, the smarty-pants will always win. The sum of the fractions will head toward infinity. We mentioned this before in Chapter 11 when we talked about things growing slowly, but we weren't in a position to justify it.

Here's what happens if we try to win the target test. Even if we hit the target for a while, we'll eventually start missing again: as we add more fractions on, the total keeps growing enough that we'll eventually fall off the other end of the target. It heads toward infinity, despite the fact that the amount we're adding on each time seems to become negligible. Note that the fact that the total keeps growing doesn't automatically mean we'll fail the target test. After all, when we add $\frac{1}{2}$ and then $\frac{1}{4}$ and then $\frac{1}{8}$ and so on, the total is also growing. It's just growing slowly enough that we never make it past 1. Whereas with the harmonic series, it's still growing slowly, but fast enough that no matter what boundary we try to impose, we will always end up going past that boundary.

Here are a few examples.

* Let's see if we can make it past 1, unlike the cake example when we eat half the leftovers each time and never quite make it to eating the whole cake until "forever," when we just about get there. So we're trying to see if we can add up enough of the fractions in the harmonic series and get something bigger than 1. And we can:

$$\frac{1}{2}+\frac{1}{3}+\frac{1}{4}=\frac{6}{12}+\frac{4}{12}+\frac{3}{12}$$

$$=\frac{13}{12}$$

which is bigger than 1.

* Let's see if we can make it past 2. At this point I admit to resorting to a spreadsheet to work this out. My spreadsheet tells me that once I get to $\frac{1}{11}$ I'll be past 2, that is,

$$\frac{1}{2}+\frac{1}{3}+\frac{1}{4}+\frac{1}{5}+\frac{1}{6}+\frac{1}{7}+\frac{1}{8}+\frac{1}{9}+\frac{1}{10}+\frac{1}{11}>2.$$

* Now that I've resorted to the spreadsheet, let's see how long it takes to get past 5. It gets really slow but does eventually happen at $\frac{1}{227}$.

I originally thought I'd see how long it takes to get past 10, but it was taking much too long so I gave up. Clearly a better method is needed if we're going to show that this heads toward infinity. For a start, "proof by spreadsheet" is not really a valid mathematical technique. But even if it were, a method of showing that something heads to infinity isn't very good if it takes too long even to show it's bigger than 10.

Here's a much sneakier method. It's sneaky, but it's also rather clever. We clump several of the terms together in groups of one, then two, then four, then eight, and so on.

1. The first clump of fractions just has $\frac{1}{2}$ in it, the first thing in the harmonic series.
2. The second clump has the next two fractions, $\frac{1}{3}$ and $\frac{1}{4}$.
3. The third clump has twice as many fractions as the previous clump: the next four fractions, that is,

$$\frac{1}{5}, \frac{1}{6}, \frac{1}{7}, \frac{1}{8}.$$

* The fourth clump has yet twice as many fractions:

$$\frac{1}{9}, \frac{1}{10}, \frac{1}{11}, \frac{1}{12}, \frac{1}{13}, \frac{1}{14}, \frac{1}{15}, \frac{1}{16}.$$

So to sum up, we have clumps like this:

$$\underbrace{\frac{1}{2}}_{\text{1st clump}} + \underbrace{\frac{1}{3}+\frac{1}{4}}_{\text{2nd clump}} + \underbrace{\frac{1}{5}+\frac{1}{6}+\frac{1}{7}+\frac{1}{8}}_{\text{3rd clump}} + \underbrace{\frac{1}{9}+\frac{1}{10}+\frac{1}{11}+\frac{1}{12}+\frac{1}{13}+\cdots}_{\text{4th clump}}$$

Now let's pause for a second and take stock. I'm going to show that each of these clumps adds up to *more than* $\frac{1}{2}$, apart from the first one, which only has $\frac{1}{2}$ so adds up to exactly that.

Because we're lazy (or rather, because I like conserving brainpower), I'm not actually going to add these clumps up – I'm just going to show that they add up to something more than $\frac{1}{2}$. Take the second clump, the one containing $\frac{1}{3}$ and $\frac{1}{4}$. Each of these numbers is greater than or equal to $\frac{1}{4}$. So the total is greater than or equal to $2 \times \frac{1}{4}$, because there are two terms, each greater than or equal to $\frac{1}{4}$. But $2 \times \frac{1}{4} = \frac{1}{2}$.

$$\underbrace{\frac{1}{3}+\frac{1}{4}}_{>\frac{1}{4}+\frac{1}{4}=\frac{1}{2}}$$

That was a bit of an over-the-top way of showing that $\frac{1}{3} + \frac{1}{4} > \frac{1}{2}$, instead of just adding the fractions up. However, the point is we're coming up with a method here that will *generalize* and enable us to get arbitrarily far along the sum, whereas putting the fractions over a common denominator to add them up, or using a spreadsheet, did not generalize well. Techniques that generalize well often look silly when you apply them to basic examples. It's one of the reasons that learning math is hard, because often all the basic examples make the math look silly or pointless or contrived.

Let's keep going and take the third clump. This contains 4 terms. The smallest of the 4 terms is the last one, $\frac{1}{8}$. So all the other terms are bigger than this, which means that the sum is greater than $4 \times \frac{1}{8} = \frac{1}{2}$.

$$\underbrace{\frac{1}{5}+\frac{1}{6}+\frac{1}{7}+\frac{1}{8}}_{>\frac{1}{8}+\frac{1}{8}+\frac{1}{8}+\frac{1}{8}=\frac{4}{8}=\frac{1}{2}}$$

Now look at the fourth clump. This contains *eight* terms, and the smallest one is $\frac{1}{16}$. So the total is greater than $8 \times \frac{1}{16} = \frac{1}{2}$.

$$\underbrace{\frac{1}{9}+\frac{1}{10}+\frac{1}{11}+\frac{1}{12}+\frac{1}{13}+\frac{1}{14}+\frac{1}{15}+\frac{1}{16}}_{>\frac{1}{16}+\frac{1}{16}+\frac{1}{16}+\frac{1}{16}+\frac{1}{16}+\frac{1}{16}+\frac{1}{16}+\frac{1}{16}=\frac{8}{16}=\frac{1}{2}}$$

We can go on like this forever. The next time we'll take 16 terms, ending at $\frac{1}{32}$. Each term is greater than or equal to $\frac{1}{32}$, so the total is greater than $\frac{16}{32} = \frac{1}{2}$. Then we take 32 terms ending at $\frac{1}{64}$. Then we take 64 terms ending at $\frac{1}{128}$. And so on.

A better way than saying "keep going forever" or "and so on" is to say what happens for the *n*th clump, where n could be anything. The nth clump has 2^{n-1} terms and ends at $\frac{1}{2^n}$. Each term is greater than or equal to $\frac{1}{2^n}$ and so the total is greater than $\frac{2^{n-1}}{2^n} = \frac{1}{2}$, as we claimed.

So we see that if we add all the terms together, forever, it will be more than what we get if we keep adding $\frac{1}{2}$ to itself forever. And if we do that, we will definitely never stop growing.

Here's another way of thinking about it. If the evil opponent gives us a target and we manage to hit it, we are doomed to fall off again. Because we will eventually have added on enough further clumps of numbers, each of which is more than $\frac{1}{2}$, that our total will get too big and fall off the other side of the target.

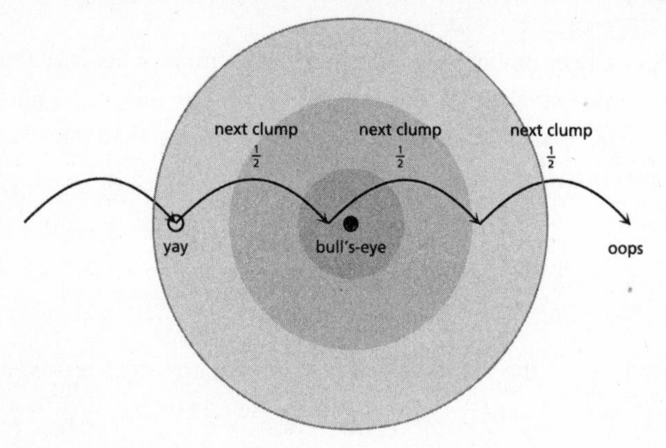

In this case, the evil opponent wins.

Bar Charts

We can draw the harmonic series as a sort of bar chart, like this:

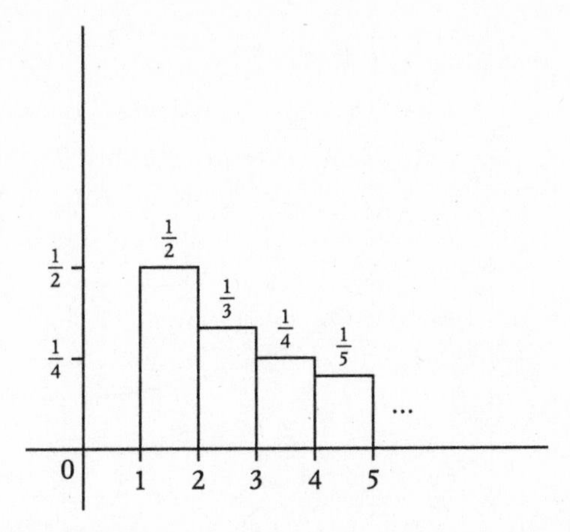

Here each bar has width 1, so the area of each rectangle is the width times the height, which is $\frac{1}{n}$ for each n. So the total area of all those bars is the same as the thing we were just trying to do, adding up all the terms in the harmonic series. This means that the area represented here is "infinite." What this really means is that no matter how big a number we think of, if we go far enough down the graph (bigger and bigger n), we can find a place where the area up to that point is definitely bigger than the number we thought of. This means that as we include bigger and bigger n, the area approaches infinity, in the sense of Chapter 11.

We are now starting to come back to the idea of the area contained by a curve. If we draw a graph of $\frac{1}{x}$ it looks like this:

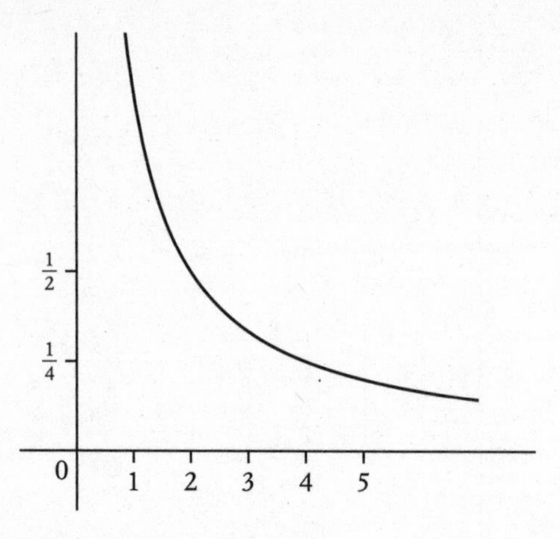

First note that $\frac{1}{x}$ gets very large when x is close to 0. We can't say where it is when x actually equals 0, but we can say that $\frac{1}{x}$ tends to infinity as x tends to 0, which means, as above, that no matter how big a number we think of, we can find a small x where $\frac{1}{x}$ is bigger.

On the other hand, as the graph goes off to the right, x becomes infinitely large and $\frac{1}{x}$ becomes infinitely small. So you might think that if we ignore the beginning, infinite part of the graph, the area under the graph should be finite. But it isn't.

If we draw this curve superimposed on the bar chart, we get this picture with the bars fitting snugly underneath:

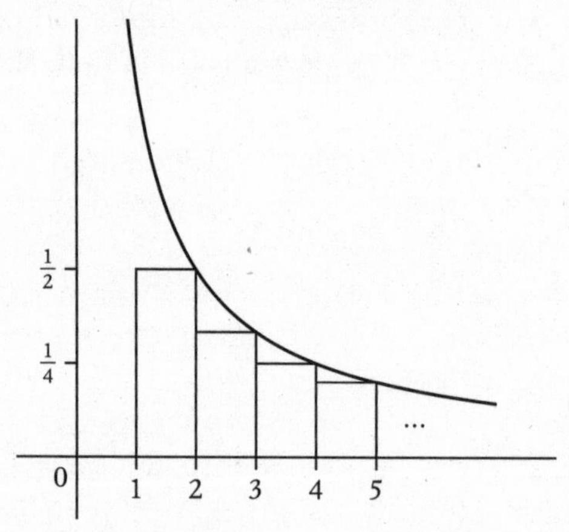

We know that the bars have a total area that is infinite, and if you look at the gaps between the bars and the curve, you can see that the bars have not even filled in all the area – the area under the curve is *more*. The area under the curve is more than something that was already infinite, so must also be infinite.

In fact, the area under this curve is measured by the natural log, written ln. (This is the log to base *e*, but we won't go into that now.) Actually it is one way of *defining* the natural log function: ln *b* is the area under this graph from 1 up to *b*. The bar chart version shows us that the natural log grows (slowly) to infinity, as we asserted in Chapter 11.

Areas Under Curves

This way of fitting a "bar chart" snugly under a graph is at the heart of finding the area under a curve. This is otherwise known as "integration." It's a bit like in the previous chapter where we chopped up a circle into neat polygons and hence triangles in order to estimate its area. Now we're going to chop the graph up into rectangular bars and use that to estimate the area by adding up rectangles. The narrower the rectangles are, the more accurate the estimate of the area should be.

Finding the area under a curve was another of the great motivations behind understanding the whole business of infinitesimally small things. At first people were trying to use the notion of an actual "infinitesimally small" length, and trying to say what it would mean to have rectangles that had "infinitesimally small" width. Mathematicians were worried about this approach, and rightly so – as we've said before, an infinitesimally small length isn't something that makes sense mathematically as a real number, any more than an infinitely large real number does.

Instead, it was another of those moments where two people worked out how to do it at similar times, in different countries, in different ways. This time it was Lebesgue and Riemann, but it wasn't quite as close together as Cantor and Dedekind constructing the real numbers. Riemann did it using the vertical bars just like we had in the example of $\frac{1}{x}$ above. Curiously, Lebesgue did it using horizontal bars. (This is an oversimplification but gives the general idea of the difference.) Lebesgue's way is perhaps less intuitive but more powerful. I won't go into it here as it requires a lot more technicality.

Riemann's way was to use vertical bars to give two estimates of the area: an overestimate and an underestimate. Take $\frac{1}{x}$ again, for example. If we just chopped it up into vertical bars,

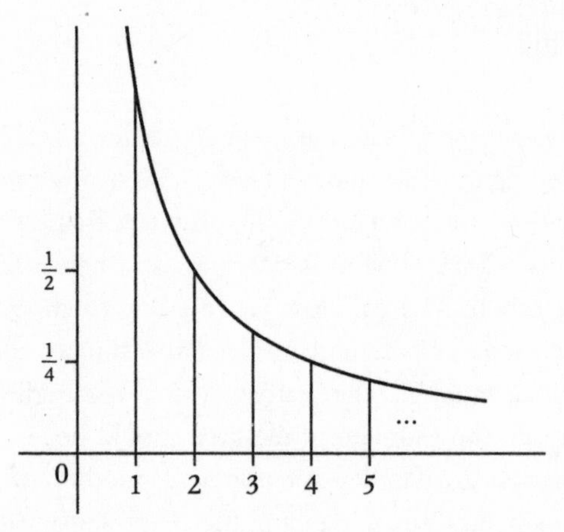

there are two ways we could make rectangles. We could put the horizontal lines either above the graph

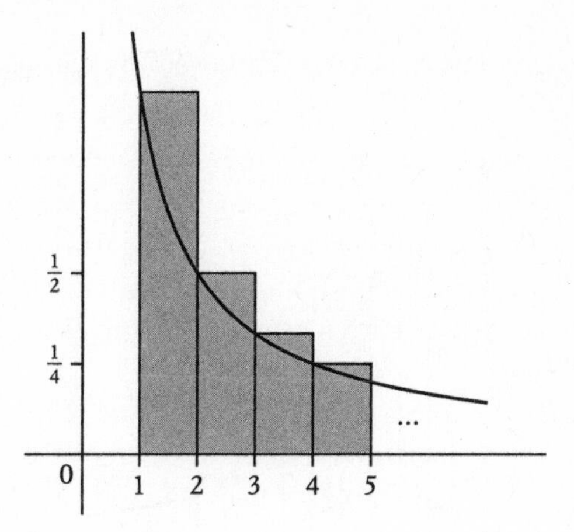

or below the graph.

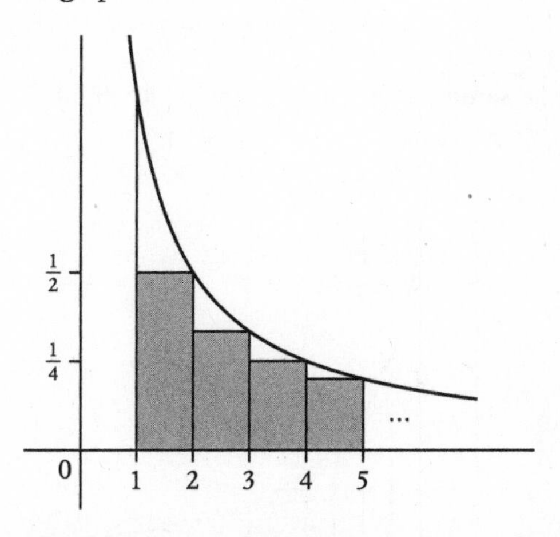

This is a bit like when we approximated the circle with a polygon fitting outside the circle and also with a polygon fitting inside the circle, with the circle sandwiched in between. Here if we put the rectangles above the graph, then we've included *more* space than the graph has under it, so it's an overestimate of the area under the graph. But if we put them below, then we've omitted some space that the graph has under it, so it's an underestimate. If we make the bars narrower, both of those estimates will get closer to the curve and so improve. For example, if we make the bars 0.5 wide instead of 1, we get this, and the overestimate area has "improved" by the shaded parts shown below:

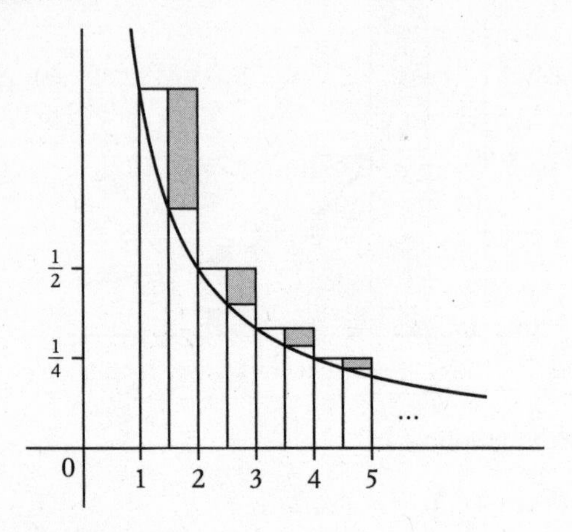

If we do the same for the underestimate, it will improve by the following shaded parts:

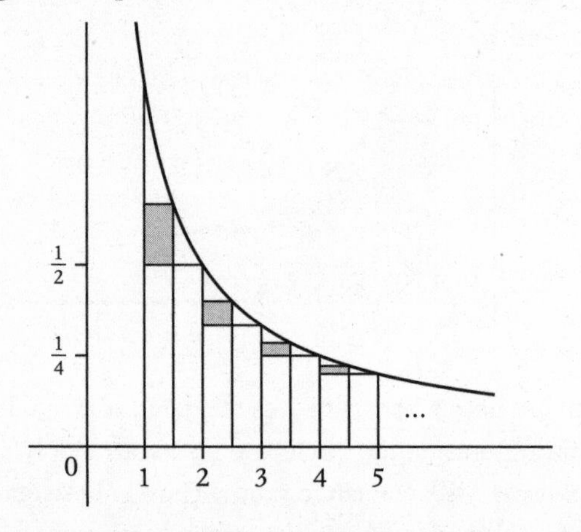

The genuine graph is sandwiched between. Now the question is whether the underestimate and the overestimate ever meet. They might not ever *actually* meet, but they could get so close that the distance between them has become negligible. How negligible? Negligible enough to defeat the old smarty-pants again. It's as if you and your friend are doing the target test together, and you have to be able to *both* hit the same target forever, no matter how small the target is. If you can both do it together, then you win. Again you don't necessarily know where the bull's-eye is, but there must be one somewhere, and that is the answer to the area under the curve.

When you learn integration at school, you usually learn it as some kind of nifty formula so that if you're asked to integrate a particular function, you manipulate the function and know what the answer is. But the *reason* that works is this process of chopping the graph into infinitesimally small bars. Just like with the infinite cookies, the target test is how we make mathematical sense of what that means.

The Ultimate Cookie Conundrum

We'll now go back to that last, most weird, cookie conundrum, where we had an infinite number of cookies made from a finite amount of dough, and yet the cookies stretched out to infinity when we lined them up. How is that possible?

We were making cookies where the second one had radius $\frac{1}{2}$ the first one, then the third had radius $\frac{1}{3}$ the first one, then $\frac{1}{4}$, and so on.

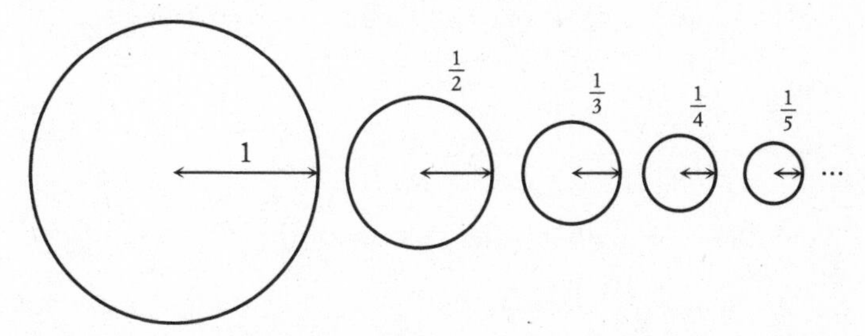

When we line them up, the distance they take up is twice the harmonic series, that is, double something that was already infinite. So it's definitely infinite. But what about the volume of cookie dough? Why is the volume of cookie dough still finite?

Let's assume the cookies are circular and perfectly even. I was once criticized in public for assuming this about scones, and accused of using factory-made scones. For the record, I do make my own scones, and I'm perfectly aware that nothing on earth is *perfectly* round and even, but it's a good enough approximation for a math discussion! It's not exactly a life-and-death situation, after all.

Anyway, the volume of cookie dough for each cookie will be the area of the circle times the thickness of the cookie. The area is πr^2 and the radius of the nth cookie is $\frac{1}{n}$. Let's call the thickness some fixed amount t. It's a bit odd to imagine the thickness staying the same even as the cookies become infinitesimally tiny, but this still gives us a finite total volume, so if the thickness became smaller as the cookies did, we would *definitely* have a finite total volume.

This way of making slightly unrealistic assumptions is key to pure math. We're trying to answer a particular question, and so we make various assumptions along the way if we think it doesn't significantly affect the answer to the question, even if it makes the calculation less "lifelike." If we were trying to answer the question "Exactly how much dough do we use?" then the assumption that the cookies are all the same thickness would certainly affect the answer. However, we're only trying to answer the question "Is the amount of dough finite or infinite?"

Given these assumptions, the volume of the nth cookie is πr^2 times the thickness, that is,

$$\pi \times \left(\frac{1}{n}\right)^2 \times t = \frac{\pi t}{n^2}$$

which gives us this sequence of volumes:

$$\frac{\pi t}{2^2}, \frac{\pi t}{3^2}, \frac{\pi t}{4^2}, \frac{\pi t}{5^2}, \frac{\pi t}{6^2}, \ldots$$

Now the πt part doesn't ever change, so let's pause and think about the only part that does change, that is, the numbers

$$\frac{1}{2^2}, \frac{1}{3^2}, \frac{1}{4^2}, \frac{1}{5^2}, \frac{1}{6^2}, \ldots$$

This is like the harmonic series

$$\frac{1}{2}, \frac{1}{3}, \frac{1}{4}, \frac{1}{5}, \frac{1}{6}, \ldots$$

but now each term is squared. (It's also what we'd get if we made square cookies instead of round ones.) Squaring a fraction that's smaller than 1 makes it even smaller (because you're taking a fraction of a fraction), so these numbers get smaller *faster* than the ordinary ones in the harmonic series. This turns out to be key to knowing whether a sum grows to infinity or stabilizes to a limit. The individual terms have to get smaller, but they have to do it quite fast. The ones in the harmonic series don't do it fast enough. The ones in the squared version *do* do it fast enough. So this sum is finite. In fact, the target it lands on has its bull's-eye at $\frac{\pi^2}{6} - 1$.

This is hard to prove, but you can try and convince yourself it's true by trying to arrange small squares inside a 1×1 square. If we use a small square whose edge is $\frac{1}{n}$ long, then the area of the small square will be $\frac{1}{n^2}$. So filling in the 1×1 square with squares whose edges have lengths

$$\frac{1}{2}, \frac{1}{3}, \frac{1}{4}, \ldots$$

is like finding the sum

$$\frac{1}{2^2} + \frac{1}{3^2} + \frac{1}{4^2} + \cdots$$

and comparing it with the area of the big square, which is 1.

We can start by putting a $\frac{1}{2} \times \frac{1}{2}$ square in the corner of the 1×1 square, say like this:

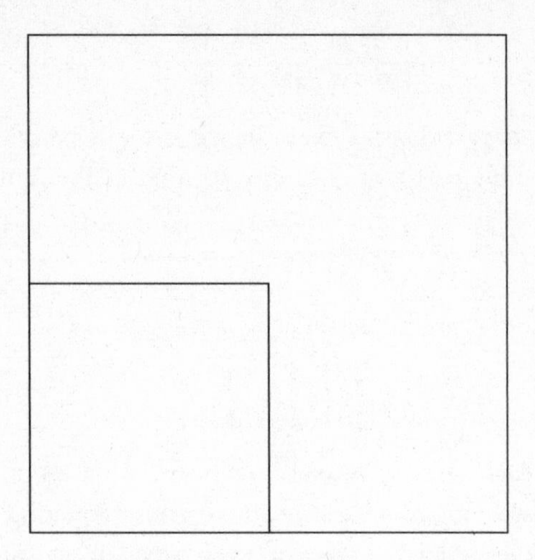

Next we have to put in a $\frac{1}{3} \times \frac{1}{3}$ square. We could put it here:

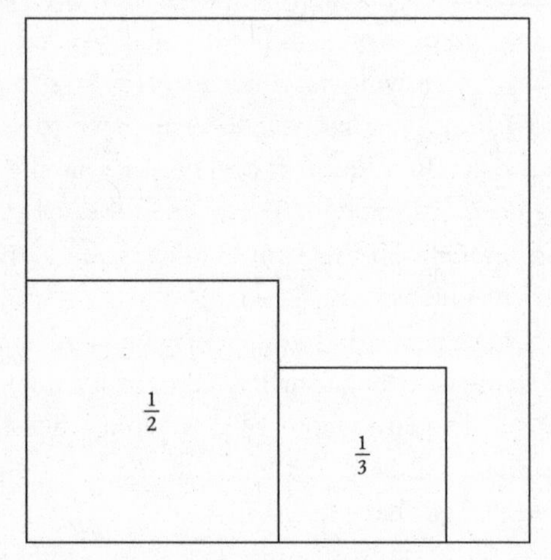

For the $\frac{1}{4} \times \frac{1}{4}$ square we can't put it in the bottom right corner now because there's not enough space left (the remaining space is only $\frac{1}{6}$ wide), but we can put it above the $\frac{1}{2} \times \frac{1}{2}$ square like this:

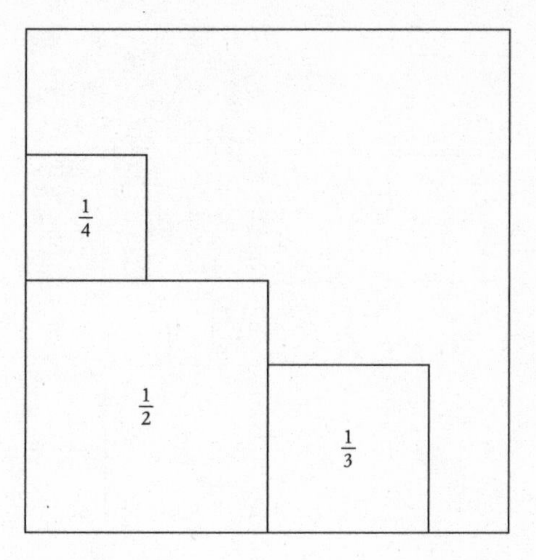

We could keep going for a few more steps as follows:

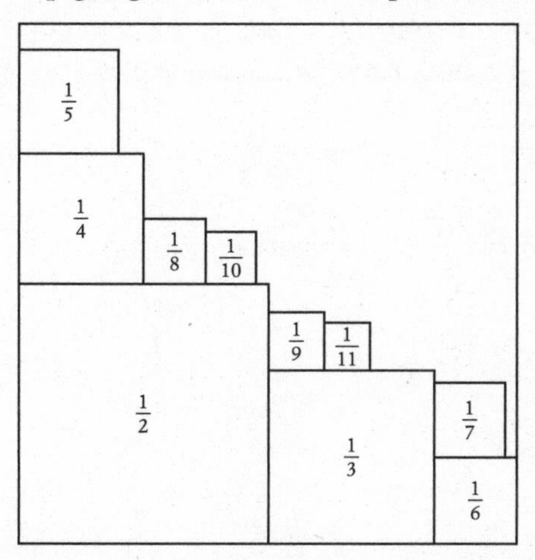

If you try it for yourself, you'll find that there's always a huge amount of space left for you to position the next square. This isn't a proof, but it certainly *feels* like you'll never run out of space. In fact, you might agree it feels like you won't even need the top right-hand quarter of the square:

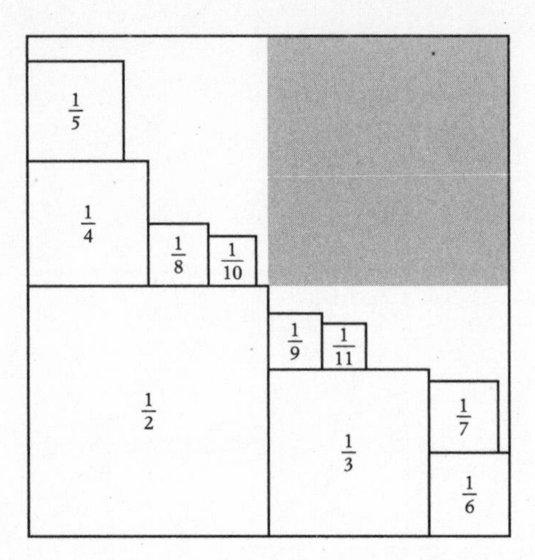

meaning that we can fit all these small squares into the area of $\frac{3}{4}$. This is right – if you try putting $\frac{\pi^2}{6} - 1$ into your calculator, you'll find it's about 0.645, which is less than $\frac{3}{4}$.

If you can remember any integration, it's not too hard to *prove* that the sum is finite, by thinking about the graph of $\frac{1}{x^2}$.

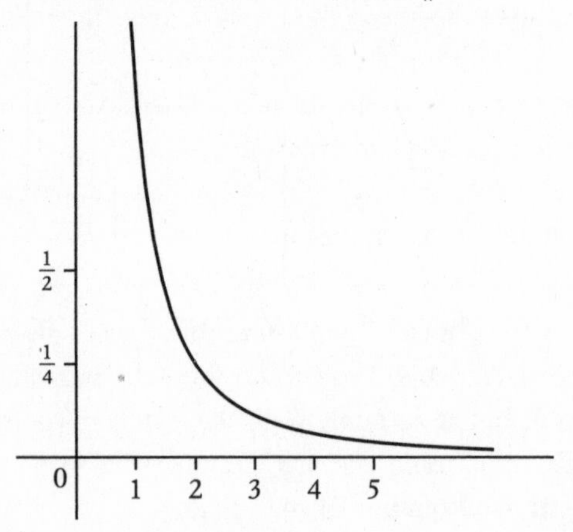

The area under this curve will have to be more than the sum we're looking for, if we think about chopping it up into vertical bars like this.

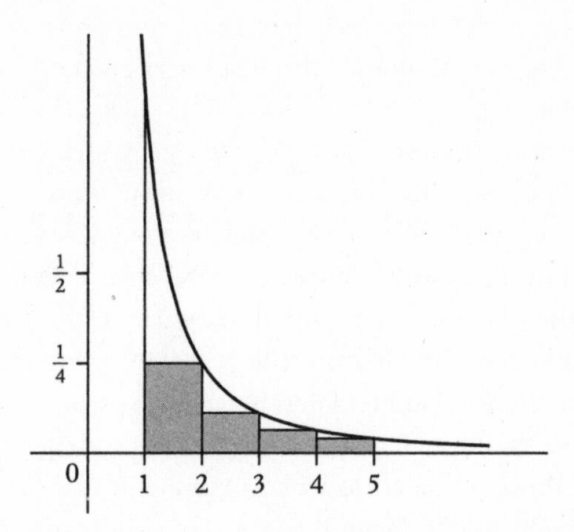

So if the area under this graph is finite, then our sum must be finite too. If you remember how to integrate $\frac{1}{x^2}$, it's not very hard to work out that the area under the curve from 1 to b is $1 - \frac{1}{b}$, which is less than 1 no matter how big b becomes. So it's definitely finite.

Now, the amount of cookies that we have isn't exactly the same as this "squared" version of the harmonic series, but we just have to multiply the whole thing by π times the thickness of the cookies. And because it was finite before that, it must still be finite afterward. (Unless we have infinitely thick cookies. Sadly, we don't. Or maybe that's a good thing, otherwise I'd be infinitely fat.) Another way of thinking about this final step is that we could fit each of our round cookies inside each corresponding square in the picture, with space to spare.

So the total *volume* of cookie dough is finite, even though when we line all the cookies up side by side, they stretch an infinite distance.

Implausible Volumes

This weird situation with the cookies is related to a very strange and slightly more mathematical example: the fact that you can take an infinite area, rotate it through the air, and still only have swept out a finite volume. The idea of rotating a shape through the air to make a volume is a bit like using a potter's wheel. The wheel rotates, so whatever shape you make with your hands ends up being made all the way round the pot. I've always been fascinated by these but have never had a chance to use one. It reminds me of the magic of cutting snowflakes by folding paper up and making one cut turn into 8, or 16, symmetrically. A potter's wheel is a kind of smooth version of that.

Imagine taking one of those big bubble-making hoops that you drag through the air to make a bubble. If you drag it in an entire circle, you'll have made a doughnut shape (a ring doughnut), formally called a *torus*. You could imagine taking other shapes of a bubble-making hoop and waving them through the air in a big circle. You'd make something that had circles in one cross section and hoop shapes in the other. You might have seen those Chinese paper fold-up toys, where you unfurl the shape in a fan-like fashion and complete a circle with it, so you make a little creature that's slightly "solid."

We could do it with this shape:

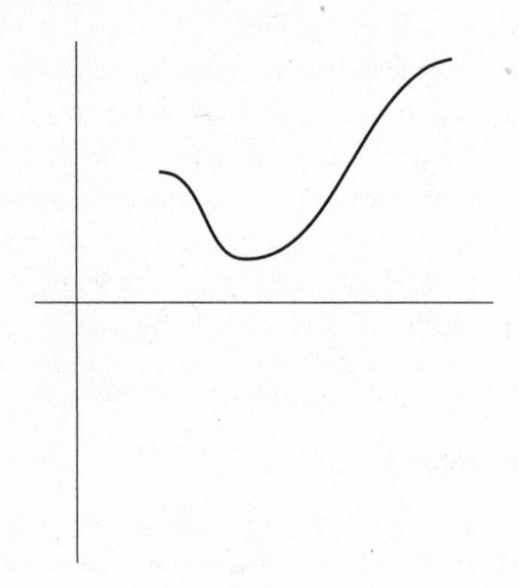

and rotate it all the way around the x-axis to make something a bit vase-like:

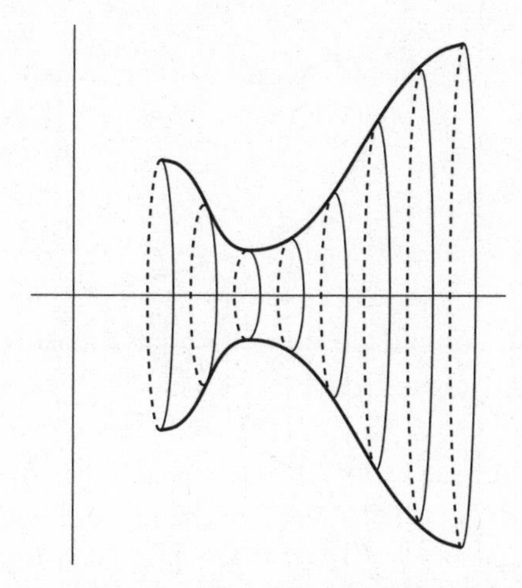

These are called "volumes of revolution," because they're made by revolving through one complete revolution about the axis.

Now if we take the graph of $\frac{1}{x}$ and rotate it around the x-axis we get this:

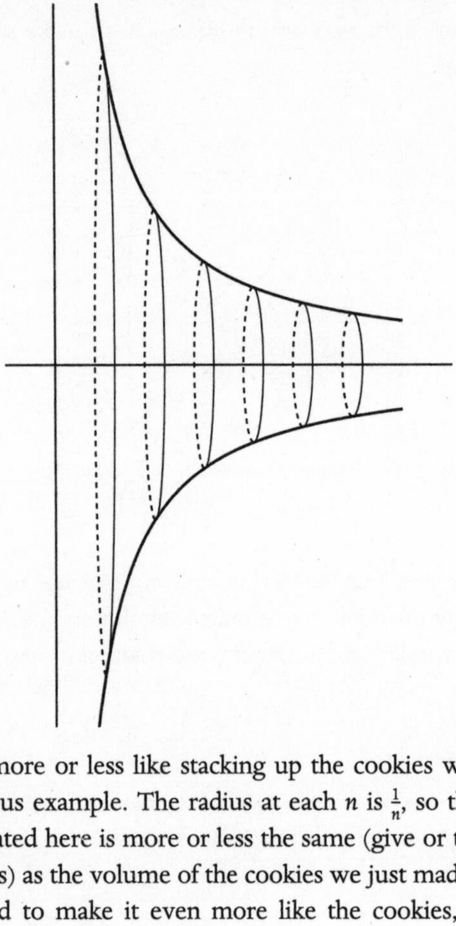

which is more or less like stacking up the cookies we made in the previous example. The radius at each n is $\frac{1}{n}$, so the volume we've created here is more or less the same (give or take a little curvedness) as the volume of the cookies we just made. In fact if we wanted to make it even more like the cookies, we could rotate these rectangular bars instead:

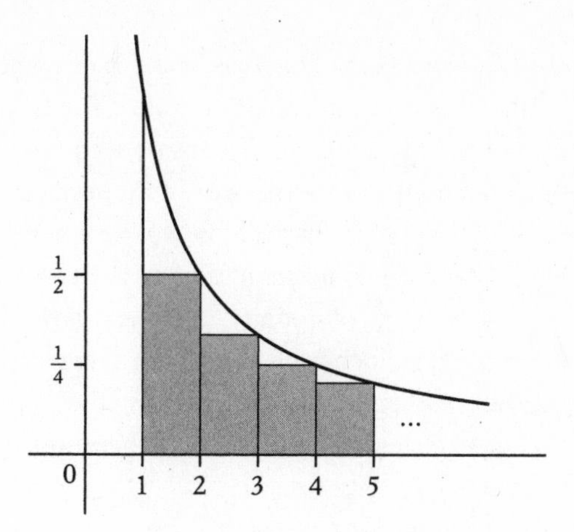

to give this.

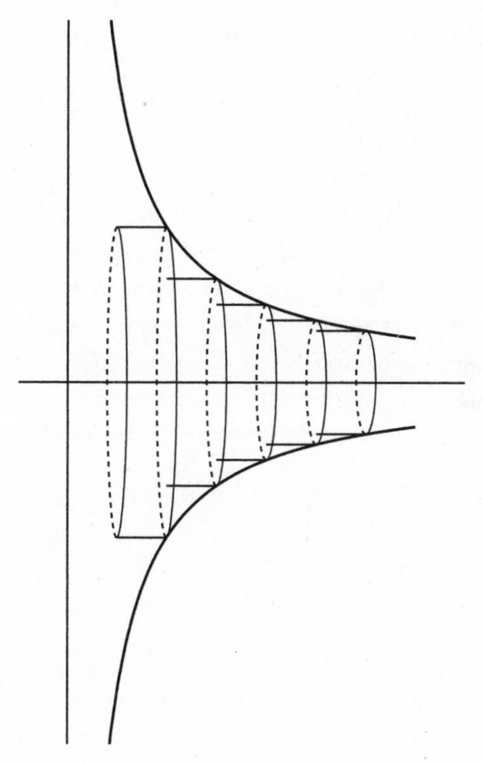

Now here's the weird part. The *area* of the rectangular bars is infinite. But the *volume* of the cookies is finite. The height of the infinite stack of cookies is infinite. The cross section of the stack of cookies is infinite! (It's twice the area of the rectangular bars.)

I hope you're feeling disconcerted now, like you're on a boat that's rocking up and down in the middle of a vast, crazy ocean. That's fine. It's the joy of infinity. The real joy, to me, is understanding the logic behind it, and seeing how to deal with these things with mathematical rigor. Because then you can ride those waves, not get seasick, and even, perhaps, feel positively exhilarated by it.

I recently went to Sydney for a conference, and the day before it started I went on a whale-watching boat trip. The sea was quite rough, but I remembered to ride with the ups and downs of the boat and not try to resist it. By the time we were far out of the harbor, almost everyone else on the boat was sick and had to lie down, and only a few of us could really enjoy the incredible power, majesty, and elegance of the humpback whales. This is what math can feel like as well. It's a very different kind of power, majesty, and elegance that we're going out looking for, but it's still a pity that so many people get seasick on the way there.

17 Where Infinity Is

My favorite Winnie-the-Pooh story is the one where Pooh and Piglet are tracking some footprints round a tree. They think they're following a heffalump. When they get back to where they started, they don't realize it's their own footprints in the snow that they can see, so they think that the first "heffalump" (who was in reality Pooh) has now been joined by two more. Except that the two extra sets of footprints in reality belong to Pooh (going round the tree the second time) and Piglet, who has now joined him in the hunt. Only after they've been round a couple more times does it dawn on them what is going on, and they sheepishly go home.

As I mentioned at the start, I grew up in a house with a fireplace and chimney in the middle. We did have central heating, but the original idea was that the chimney would heat up the center of the house, and every room in the house was clustered around it. This meant that instead of the house having an entrance with rooms coming off it, the rooms led onto each other round the chimney.

The best thing about this when I was little was that my sister and I could chase each other round the house in circles, looping round and round the fireplace, through the kitchen, dining room, sitting room, kitchen, dining room, sitting room. Then of course one of us would suddenly change direction and try to catch the other "round the back," and then we'd all scream and run round in the opposite direction – kitchen, sitting room, dining room, kitchen, sitting room, dining room. The wonderful thing about houses with this sort of circular loop is that you essentially have an *infinite* house. That is, there's an infinite path you can walk along. It's true that you will keep coming back to the same place, but there's something different about coming back to the same place having gone round something, as opposed to just going from *A* to *B* and back to *A* again. If you were like Theseus in the labyrinth, trailing a piece of string behind you, then going round the chimney back to where you started is definitely different from just going from *A* to *B* and back, because the string will really have done something and will now be looped around the chimney. You've been on a real journey.

Mathematics deals with this phenomenon in the form of something called "covering spaces." The idea is that instead of drawing a map of a building showing where everything is, you draw a map showing all the paths that you can take, trailing Theseus's string behind you as if in a labyrinth. Imagine tying one end of the string to the entrance and tying the other end of the string to you so that nothing becomes detached when you plunge into the labyrinth. If you retrace your steps back out, your string will come with you. But if you take a very different route out, the string might be stuck inside, looped through the tunnels. If I went back to my childhood home and attached some string to the front door and then attempted to run round and round the chimney place, I'd need an infinitely long piece of string. (And a bodyguard to stop me being taken away and locked up.) In the mathematical covering-space interpretation this would turn into a path that was actually infinitely long, instead of looping back on itself repeatedly. It would be as if, like Piglet and Pooh, you don't realize each time you get back to where you started.

One of my ambitions is to live in a house with a genuine circular loop in it again. Among mathematicians I call this a house "with homotopy," because homotopy is the mathematical concept that measures loops in spaces. (We mentioned this in Chapter 13.) If it's the kind of loop that you could just pull closed with your Theseus-like trail of string without it getting stuck on something like a chimney, for example, then it doesn't count as a real loop. It only counts as a real loop if your string will get stuck round it. Not like after I had my knee operation and the longest walk I could take was a sad and pointless little circle around my kitchen.

I am particularly intrigued by houses with more than one staircase. Apart from them being very grand, I like the fact that this gives them a *vertical* loop, as you can go up one staircase, go down the other staircase, and go back to the bottom of the first staircase. You would not be able to pull your Theseus-like trail of string closed.

Where I currently live, I have no real loops. I have no homotopy.

Infinite Paths

Even though I don't often chase my sister round in circles anymore, I still like buildings with loops in them because it means I can go on an infinitely long walk. Mathematicians – and no doubt others – often like walking to help them think. There's something about the gentle movement that focuses my thoughts. Sometimes I think it's that the act of walking keeps the logistical parts of my brain just busy enough that my dreamy mathematical brain can roam free without the other part constantly bothering it and saying things like "Don't forget to buy some eggs."

However, the trouble with going for an infinitely long walk is that usually you would end up infinitely far away from home. Also, if I go for a walk outdoors and think about math at the same time, I am very likely to get lost and also be hit by a car.

Having a building with a loop in it solves all these problems, because it means you can walk round and round in circles "forever," without feeling like the circles are futile, because unlike when I just walked in a circle around my kitchen, you're really going round something solid, not just empty space. As I have already mentioned, the math department at the University of Nice is entirely based on a circular corridor. The inside of the circle is a beautiful circular inner courtyard, with floor-to-ceiling windows from the corridor looking down on it. All the offices are on the outside of the circle, and as you walk round and round the corridor you are always facing onto this beautiful courtyard. I'm glad to say I was not the only person who sometimes just walked round this in circles, thinking. There's something about the possibility of infinity that is liberating.

The problem with this department was it was so symmetrical it was impossible to remember where I was. There were four staircases, equally spaced around the circle, and whenever I emerged from a staircase I could never work out whether it would be quicker to go left or right to get where I wanted. The funny thing is the number of times I ended up going round the circle *more than once* to get where I wanted, because I missed it on the way.

Paternoster

Another satisfyingly infinite structure is a paternoster. There is a quite famous one of these in the Arts Tower at the University of Sheffield. It's like an elevator, except much more interesting, consisting of a circular loop of capsules, each of which carries two people (in Sheffield's case). At any given moment there's one capsule at each floor going up, and one going down, and a couple in the process of looping over the top or under the bottom. The paternoster has no doors, and it never stops moving. You just have to jump into a capsule as it goes past where you're standing, and jump off when it reaches the floor of your destination. It moves quite slowly in reality, but still the first few times I took it I was terrified, and my pulse shot up when I had to get on and off. It's amazing that this thing hasn't fallen foul of health and safety regulations yet. It is technically forbidden to ride over the top or under the bottom, but it is everyone's first thought, and I'd guess there aren't many people at the university who haven't tried it.

It's called a paternoster because it's supposed to remind us of a string of rosary beads, and the idea of rosary beads is to keep track of how many times you've said certain prayers. There might be ten small beads in a row to keep count of saying ten Hail Marys, and then a larger bead to indicate that it's time to say a Lord's Prayer before starting another ten Hail Marys. *"Pater noster"* is how the Lord's Prayer starts in Latin. The idea of the beads is that having a physical way of keeping track of the number of repetitions leaves your mind more free for thought.

I feel similarly about paternosters. I wish there were always paternosters in buildings instead of elevators. To me it makes the whole building feel connected in an infinite loop, rather than being made up of separate, finite-sized floors that you have to travel between. It means that my brain doesn't feel divided up into floors when I'm moving between floors, and the predictable continuous motion means that my mind can feel more free for thought than when I'm waiting for the unpredictable arrival of an elevator. This is a bit melodramatic, but when I'm thinking about math, this sort of thing makes a surprising amount of difference to my progress.

Besides, a paternoster is an amazing vertical version of being able to go round and round in circles (although you can't really chase anyone round it), and it makes me feel that a building is infinitely big. If you attach your Theseus-like string to the ground floor and then get in a paternoster and go round and round in it, you will again need an infinitely long piece of string.

Other Circles

We previously claimed that there was nothing really infinite in the world, but now we see that circles are wonderfully infinite if you think about the paths that you can take on them. Circular (or oval) racetracks are clever because you can run any length of race on them. A little boy called Jacob recently wrote to me to tell me that math is his favorite subject, and also, by the way, that he could do 84 meters on the monkey bars. I wondered if he had a very long series of monkey bars, or kept turning round, or if he had circular monkey bars so that he could keep going just as long as he wanted.

The Circle Line on the Tube used to be more fun because you could sit on the train forever and keep going round and round, which is much more like an infinite journey than going back and forth between the end stations of a noncircular line. Some people are very excited by driving round and round the M25 around London repeatedly (though personally I'm always glad if I just manage to get partway round it without getting completely stuck in traffic). I expect there are also such fanatics of other circular highways round cities, like the Périphérique around Paris, or the Capital Beltway around Washington, DC.

An even more interesting version of a circle is a Möbius strip. This is made from a strip of paper where you stick the ends together, but instead of sticking them in the obvious fashion,

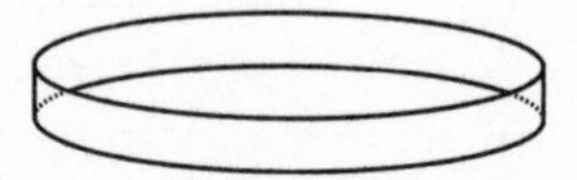

you twist the paper before sticking the ends together.

The intriguing thing about this, both physically and mathematically, is that you have now stuck the front of the paper to the back, and the back to the front, meaning that there is now only one side – the front and back have become the same.

There are many interesting things you can do with this, but one simple one is just to pick it up and go round it with your fingers. It's even easier to lose where you are on it than if you're just going round a circle, as you can get confused about whether you're on the front or the back – because there's no difference. You can also try tracing your finger around the edge of the Möbius strip. The edge is a single circle, but it loops back on itself and then appears to "go round" the Möbius strip again, although in reality it's only gone round once. It's a sort of figure eight, another very satisfying shape to go round and round forever. How fitting that these should be the symbols that begin and end our quest for infinity:

$$0 \qquad \infty$$

Except now we know there is no end, because not only does infinity go on forever but the hierarchy of infinities goes on forever, with bigger and bigger infinities, even though we're sitting quietly inside our large but finite world. Perhaps we should no longer be surprised that something infinite can fit inside something finite. Where infinity is concerned, it seems that almost anything is possible. The world of mathematics fits in our brain but is bigger than the universe.

Infinity is liberating but sometimes *too* liberating. I can think much more freely and creatively if I feel I have an effectively infinite amount of time at my disposal. I am more likely to prove a theorem if I have no fixed engagements for the rest of the day, even if the proof only takes two hours. If I only had two hours available, I probably wouldn't be able to do it.

And yet we've seen that if we were immortal, we could procrastinate forever. Knowing me, I probably actually *would* procrastinate forever. Having infinite dimensions at our disposal gives us infinite subtleties that we can't deal with. But we can still dream about them, even if we can't explain all those dreams.

I truly believe that trying to explain everything isn't the point. Rather, the point is to explain as much as we can, and, importantly, to be clear where the boundary is between what we can and can't explain. In my mind's eye, the sphere of things we can logically explain is at the center of the universe of ideas, and the aim of mathematics is to move as much as possible into that sphere. So the sphere is always expanding, and as it does so, its surface keeps growing. The surface is where the explicable and the inexplicable meet.

The most beautiful things to me are the things just beyond that boundary of logic. It's the things we can get quite a long way toward explaining, but then in the end they just elude us. I can get quite a long way toward explaining why a certain piece of music makes me cry, but after a certain point there's something my analysis can't explain. The same goes for why looking at the sea makes me so ecstatic. Or why love is so glorious. Or why infinity is so fascinating. There are things we can't even get close to explaining, in the realm far from the logical center of our universe of ideas. But for me all the beauty is right there on that boundary. As we move more and more things into the realm of logic, the sphere of logic grows, and so its surface grows. That interface between the inside and the outside grows, and so we actually have access to more and more beauty. That, for me, is what this is all about.

In life and in mathematics there is often a trade-off between beauty and practicality, along with a contrast between dreams and reality, between the explicable and the inexplicable. Infinity is a beautiful dream, inside the beautiful dream that is mathematics.

Acknowledgments

I would like to begin by paying tribute to my late student Lisa Kuivinen, who showed me a most interesting use of Zeno's paradox in manga. I would like to thank all my students at the School of the Art Institute of Chicago and the Universities of Chicago, Sheffield, and Cambridge, as well as the children at Park Street Primary School, Cambridge; Hillsborough Primary School, Sheffield; and Francis Parker School, Chicago. I believe teaching is a two-way process, and I have learned a great deal from my students over the years.

None of this would happen without the support and inspiration of my parents, my sister, and my little nephews, Liam and Jack.

Thanks are also due to my friends, whose insights and intrigues have always spurred me to think about things differently and more lucidly. Special thanks are due to those whose thoughts I have specifically referred to in this book: Amaia Gabantxo, Jason Grunebaum, Christopher Danielson, Richard Wood, Tom Crawford, Sally Randall, David Hutchings, Sam Duplessis, Katherine Fincher, Alice Sheu, Tom Crawford, Gerald Finley. Thanks for help with transatlantic cultural references are due to Courtney Nzeribe, Timothy Madden, and Rohan Zhou-Lee.

Chapter 12 is dedicated to Gregory Peebles and his ongoing love of dimensions.

I'd like to thank my English teacher Marise Larkin for teaching me how to write essays in a way that generalized very well to books; my agent, Diane Banks; Nick Sheerin and Andrew Franklin at Profile; TJ Kelleher and Lara Heimert at

Basic Books; Sarah Gabriel for continuing to be my beacon in brain fog; Reuben Thomas for help with Ubuntu and LaTeX; Oliver Camacho for always being there.

Finally, thank you to Michael at Acanto for keeping me nourished with food, drink, and discussions about Zeno.

Index

Eugenia Cheng is a Scientist in Residence at the School of the Art Institute of Chicago and an honorary fellow of the University of Sheffield. The author of *How to Bake Pi*, she lives in Chicago, Illinois.

Photograph by Eugenia Cheng and Matt Wechsler